Selected Topics in Micro/Nano-Robotics for Biomedical Applications

Yi Guo
Editor

Selected Topics in Micro/Nano-Robotics for Biomedical Applications

Springer

Editor
Yi Guo
Department of Electrical
and Computer Engineering
Stevens Institute of Technology
Hoboken, NJ
USA

ISBN 978-1-4419-8410-4 ISBN 978-1-4419-8411-1 (eBook)
DOI 10.1007/978-1-4419-8411-1
Springer New York Heidelberg Dordrecht London

Library of Congress Control Number: 2012945728

Printed on acid-free paper

Springer is part of Springer Science+Business Media (www.springer.com)

Preface

My motivation for editing the book came from a perceived need for educational materials in the interdisciplinary subject area of micro/nanorobotics in biomedical applications. This perception emerged from two fronts. On one hand, I have been looking for new robotic techniques to introduce in my advanced robotics class at Stevens. On the other hand, the recently established Biomedical Engineering (BME) program at Stevens needs to expand its current curriculum so that cutting-edge engineering technologies supporting biomedical applications are included. During the 2005–2006 academic year, a group of four Biomedical Engineering undergraduate students initiated a so-called BioCybernetics project to develop a nano-scale autonomous device that would have the potential to diagnose and perhaps, treat illness in the field of gastrointestinal endoscopy. The project eventually involved 15 students from mechanical, electrical, chemical, computer, and biomedical engineering at all levels from freshman to seniors. In the end, the students realized that only limited objectives could be met within the resources available and within the time frame of one academic year. Several faculty members from multiple disciplines were approached by the students for guidance on various aspects of the project, and it was recognized that the group of students would have achieved much more in the BioCybernetics project if some training materials on this topic are made available to them.

After several discussions with faculty from different departments, we realized that this is not only a Stevens' problem, but also there is a national need for more educational materials on the emerging topic. Over the last decade, there has been a significant growth in the number of undergraduate BME programs and the number of students enrolled in these BME programs. Based on a recent forecast by the US Bureau of Labor Statistics, biomedical engineering jobs will climb almost twice as fast as the overall average for a 26.1 % gain by 2012 while overall job growth is projected to be 14.8 %. Micro/nanorobotics for biomedical applications is an emerging area that has received advancement during the last decade. However, contrary to the large amount of teaching and learning materials on large-scale medical robots, there is a general lack of learning materials on micro/nanorobotics in undergraduate education, and the national growth of the BME programs calls for new educational materials supporting biomedical engineering.

Motivated by the above-mentioned reasons, we have put together an educational proposal for the National Science Foundation's Course Curriculum and Laboratory Improvement Program (CCLI), which is now named as Transforming Undergraduate Education in Science (TUES). We were fortunate to get it funded. With the support, we developed case studies and laboratory modules to introduce robotics concepts to BME students. Expanding the effort, we have invited other researchers in the field to contribute to the rich interdisciplinary aspects of the subject as sampled in the chapters of the book.

Principle audience of the book includes upper-level undergraduates and graduate students in biomedical engineering and other disciplines where micro/nanorobotics techniques could be readily applied. Interested readers may also include researchers of relevant fields and biomedical engineers who are interested in robotics technologies. It is impossible to include every piece of ideas and research work on the subject. It is my wish that a representative selection of them as presented in the book will inspire a student to pursue a career in the challenging field, help to generate an awareness of the diversified technology, and provide some guidance to interested readers.

This book would not be possible without the research support from the funding agencies including National Science Foundation's DUE, CMMI, EFRI programs. I would also like to acknowledge my co-PIs on the NSF CCLI project, Profs. Arthur Ritter and Hong Man, who helped me to pilot-test some of the developed materials. Sincere thanks go to the chapter contributors of the book, who have spent their precious research time on the preparation of the chapters. I would also like to thank the Springer Editor, Brett Kurzman, for his patience and guidance on my first book project. Last but not the least, I would like to thank my family for their continuous support.

NJ, February 2012 Yi Guo

Contents

Contributors

Ronald S. Besser Department of Chemical Engineering and Materials Science, Stevens Institute of Technology, Hoboken, NJ 07030, USA

Vinod R. Challa Department of Mechanical Engineering, Stevens Institute of Technology, Hoboken, NJ 07030, USA

Xi Chen Department of Mechanical Engineering, Stevens Institute of Technology, Hoboken, NJ 07030, USA

Junjun Ding Department of Mechanical Engineering, Stevens Institute of Technology, Hoboken, NJ 07030, USA

Zaili Dong State Key Laboratory of Robotics, Shenyang Institute of Automation, Chinese Academy of Sciences, Shenyang 110016, China

Frank T. Fisher Department of Mechanical Engineering, Stevens Institute of Technology, Hoboken, NJ 07030, USA

Baofeng Gao Faculty of Engineering, Kagawa University, Kagawa, Japan

Shuxiang Guo Faculty of Engineering, Kagawa University, Kagawa, Japan

Yi Guo Department of Electrical and Computer Engineering, Stevens Institute of Technology, Hoboken, NJ 07030, USA

Songyu Hu Department of Manufacturing Engineering and Engineering Management, City University of Hong Kong, Kowloon, Hong Kong

Mi Li State Key Laboratory of Robotics, Shenyang Institute of Automation, Chinese Academy of Sciences, Shenyang 110016, China

Shuai Li Department of Electrical and Computer Engineering, Stevens Institute of Technology, Hoboken, NJ 07030, USA

Liangqing Liu State Key Laboratory of Robotics, Shenyang Institute of Automation, Chinese Academy of Sciences, Shenyang 110016, China

Hong Man Department of Electrical and Computer Engineering, Stevens Institute of Technology, Hoboken, NJ 07030, USA

Ayokunle Omosebi Department of Chemical Engineering and Materials Science, Stevens Institute of Technology, Hoboken, NJ 07030, USA

Arthur Ritter Department of Chemistry, Chemical Biology, and Biomedical Engineering, Stevens Institute of Technology, Hoboken, NJ 07030, USA

Yong Shi Department of Mechanical Engineering, Stevens Institute of Technology, Hoboken, NJ 07030, USA

Dong Sun Department of Manufacturing Engineering and Engineering Management, City University of Hong Kong, Kowloon, Hong Kong

Youhua Tan Department of Manufacturing Engineering and Engineering Management, City University of Hong Kong, Kowloon, Hong Kong

Yuechao Wang State Key Laboratory of Robotics, Shenyang Institute of Automation, Chinese Academy of Sciences, Shenyang 110016, China

Ning Xi Department of Electrical and Computer Engineering, Michigan State University, East Lansing, MI 48824, USA

Nan Xiao Faculty of Engineering, Kagawa University, Kagawa, Japan

Xiubin Xiao Department of Lymphoma, Affiliated Hospital of Military Medical Academy of Sciences, Beijing 100071, China

Guang Yang Department of Electrical and Computer Engineering, Stevens Institute of Technology, Hoboken, NJ 07030, USA

Yafeng Yin Department of Electrical and Computer Engineering, Stevens Institute of Technology, Hoboken, NJ 07030, USA

Shubo Zhang Department of Electrical and Computer Engineering, Stevens Institute of Technology, Hoboken, NJ 07030, USA

Weijing Zhang Department of Lymphoma, Affiliated Hospital of Military Medical Academy of Sciences, Beijing 100071, China

Chapter 1
Introduction

Yi Guo

1.1 Introduction

In the same way MicroElectroMechanical Systems (MEMS) technologies have provided new medical devices in the 80s, recent development in nanotechnology is enabling the manufacturing of nanobiosensors and actuators to improve cell biology interfaces and biomolecular applications. As a consequence, nanorobotics and nanomedicine have evolved from pure science fiction to a rapid growing research area which may lead to a real implementation in a few decades. According to a 1997 report by a panel of experts sponsored by the U.S. Department of Defense, nanomedicine could become a reality by the year of 2020, and "possible applications include programmable immune machines that travel through the bloodstream, supplementing the natural immune system; cell herding machines to stimulate rapid healing and tissue reconstruction; and cell repair machines to perform genetic surgery" [1]. Also, the U.S. National Institute of Health (NIH) Roadmap's new Nanomedicine Initiatives released in 2003 envisioned that "the cutting-edge area of research will begin yielding medical benefits as early as 10 years from now".

In future decades, the principle focus in medicine will shift from medical science to medical engineering, and the design of microscopic and molecular machines will be the consequent result of techniques provided from the biomedical knowledge gained in the last century [2]. Market demand for professionals with advanced degree training relevant to micro/nanorobotics and nanomedicine will be fueled by the permeation of new discoveries in the field. In this edited book, we address selected topics in the subject of micro/nanorobotics aiming at biomedical

Y. Guo (✉)
Department of Electrical and Computer Engineering,
Stevens Institute of Technology, Hoboken, NJ 07030, USA
e-mail: yguo1@stevens.edu

Y. Guo (ed.), *Selected Topics in Micro/Nano-Robotics for Biomedical Applications*,
DOI: 10.1007/978-1-4419-8411-1_1,

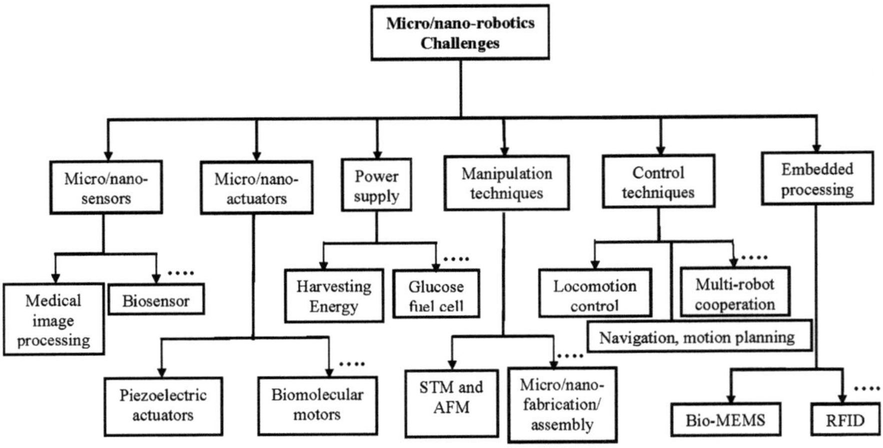

Fig. 1.1 Micro/nanorobotics challenges and supporting techniques

applications to prepare students in relevant fields with the knowledge and training to fulfill the technology demand when entering the job market.

Micro/nanorobotics for biomedical applications is an emerging area that has received advancement during the last decade. In addition to books in nanotechnology, for example, [3–5], there are growing number of articles appeared in journals and conference proceedings in biomedical micro/nanorobotics, see the survey articles [6–9] and the references therein. Among many challenges in the field of micro/nanorobotics, Fig. 1.1 shows the main topics and supporting techniques, such as micro/nano-sensors and actuators, power supply, manipulation, control, and embedded processing. Next, we briefly discuss each of these topics.

Micro/nano-sensors. In a biomedical application, the sensing of local pressure, temperature through infrared capability, proximity to surfaces through ultrasound, pH changes, and specific protein structures through functionalized surface probes would provide useful feedback data. Some of these sensors are presently being developed at the nanoscale [10, 11]. Among the relatively mature techniques, medical image processing has been used for sensory purposes in many applications, for example, the wireless capsule endoscopy. The imaging module usually includes the functions: 1) capturing images and video sequences as the robot navigates, and transmitting them to an external data recording device, 2) detecting abnormal appearance, 3) detecting important landmarks that can help the positioning and navigation subsystem. Critical constraints for the imaging module may include low computing power, low on-board memory, low energy supply, low wireless transmission bandwidth, and above all high data volume. Due to the large data volume, image and video coding is essential for effective transmission within bandwidth limited media. Wireless data transmission requires significant power consumption, and it should be kept at a minimum level as a possible. For biomedical applications, adaptive data transmission schemes need to be adopted that can provide most critical information and minimize redundant information.

Micro/nano-actuators. Piezoelectric materials have been used as actuators or sensors in bulk or thin film forms. Micron-sized piezoelectric fibers are available and have been used in commercial products. Nanoscale piezoelectric materials can generate voltages under the peristaltic action of the gastrointestinal tract and convert mechanical energy into electrical energy. This could be one solution to the power and propulsion subsystem. Another type of nano-actuators comes from the biological systems: protein motors. The classical actin-myosin power stroke that converts chemical energy of ATP to mechanical work of muscle is probably the oldest known. More recently, the microtubule motor protein families kinesins and dyneins have been identified [12, 13]. The kinesins constitute a superfamily of protein motors similar in structure to myosin that are involved in motion of the cytoskeleton of cells. The dyneins are also linear microtubule track motors which are not related to the kinesin superfamily. They are the largest of the known molecular motors and are also among the fastest. They are highly specialized for sliding the microtubules that are responsible for the beating of cilia and flagella [14–16].

Power supply is of vital importance to the design of micro/nanorobots. Existing techniques can be categorized as on-board power and transmitted power. The on-board power methods include miniature batteries, MEMS-based power generators [17] and energy harvest devices, biofuel cells [18, 19], and muscle cells [20]. Rechargeable thin-film batteries enable the use of arbitrarily shaped batteries with thickness less than 50 μm [7], and have been used in micro-robotic applications. Energy harvest devices convert various types of energy into electricity, for example, energy can be harvested from a mechanical source, such as from the environment in the form of vibration, or from other sources such as thermal gradients [21]. The transmitted power uses magnetic fields to induce electricity wirelessly, where the human body is "transparent" to magnetic fields [22, 23]. While the on-board power generation needs to consider storage methods, the wireless power transfer overcomes this limitation. But the safety of magnetic fields and the heat generated by radio frequency coils should be considered in the wireless transmitted power method.

Manipulation techniques include a board range of techniques used for micro/nanorobotics in biomedical applications, from the scanning tunneling microscopes (STMs), fabrication and assembly at the micro-scale, capturing, sorting and manipulating cells or molecules, to steerable catheters and endoscopes in surgeries. Since the inception in 1980s, STMs allow un-precedent engineering operations on molecules, atoms, and cells, which open a new era in micro/nano-manipulation and micro/nanorobotic assembly. Most recent advances in this area include nano-machines that make use of chemical synthesis, self-assembly, and supramolecule chemistry to provide building blocks for construction of complex nano-devices and nano-structures [24]. Such machines can be used to deliver drugs and targeted therapy in the human's body.

Control techniques refer to methods that control the motion and operation of micro/nanorobots in various environments, for example, navigation of pill-camera-based micro-robots in human's gastrointestinal tract [25], motion planning of steerable needles or catheters for medical procedures [26], and collective control design in nanomedicine [27]. Controlling micro/nanorobots in the human's body poses new challenges to existing control methods. Limitations of physical and environmental constraints should be considered, as well as the availability of online sensing and real-time feedback. Open-loop locomotion control has been often used to control micro-robots' motion in the human's body such as earthworm-like locomotive mechanism integrated with piezo-actuator and engraved clampers [28], and fin-moving-like motion controlled by frequency adjustment of alternate magnetic fields [29]. With the advances in nanofabrication and the availability of new tools developed in handling micor/nano-objects, unprecedented opportunity exists in breakthroughs toward precise control of micro/nanorobots for useful applications.

Embedded processing provides on-board computation and processing for micro/nanorobotic applications. MEM fabrication technology makes it possible to build structures with micron-sized features that enable embedded processing. Applied within the pharmaceutical, life sciences, and healthcare application fields, components made using this technology are referred to as "bioMEMS" devices. Key bioMEMS devices include microarrays, microfluidic chips, and micro capillary electrophoresis chips [30]. Microarray technology allows both miniaturization and high-throughput processing for applications, such as drug discovery, gene expression, genotyping, mutation screening, gene synthesis, and proteomics. While still in its infancy stage, highly integrated processing chips at the micro or nano scale could have significant impact to the field of micro/nanorobotics.

1.2 An Overview of the Book

While it is impossible to include every technique that supports the micro/nanorobotics for biomedical applications, the remaining chapters of the book represent important interdisciplinary technologies in the field. Chapters 2 and 3 present sensing technologies, which include nano-fiber techniques and biomedical imaging processing. Chapters 4 and 5 discuss power supply techniques including miniature fuel cells and energy harvesting devices. Robot navigation and control techniques are presented in Chaps. 6 and 7 using two representative applications, one on a pill-sized capsule robot operating in the human's gastrointestinal tract, and the other on cooperative drug delivery. Chapter 8 presents nanorobotic techniques in cancer targeted therapy. Chapters 9 and 10 discuss robotic manipulation techniques including optical tweezers for cell stretching and positioning, and a catheter surgery system utilizing force feedback. We highlight the contribution of each chapter in the follows.

Chapter 2 presents new sensor technology utilizing the lead zirconate titanate (PZT) nano active fiber composites to sense acoustic emission. Piezoelectric acoustic emission sensors are designed using a direct piezoelectric effect, and the sensitivity of this type of sensors is mainly determined by the piezoelectric voltage constant of the piezoelectric materials. PZT nanofibers prepared by an electrospinning process exhibit an extremely high piezoelectric constant, high bending flexibility, and high mechanical strength. Comparing to the PZT microfiber, PZT nanofibers can produce much higher voltage thus result in a higher sensitivity of the sensor. Comparing to the bulk or micro size PZT ceramic materials, the high flexibility allows PZT nanofibers to bend freely in a transverse direction, thus the sensor can be easily bonded to curved structures. The PZT nanofiber acoustic emission sensor can also achieve self-powered active sensing. The authors of Chap. 2 develop the PZT acoustic emission sensor that consists of nanoscale PZT fibers with interdigitated electrodes on a silicon substrate, and present the design, fabrication, and characteristics (including elastic wave detection, signal attenuation, and acoustic emission detection) of the developed sensor in this chapter.

Chapter 3 presents biomedical imaging processing techniques for the wireless capsule endoscopy (WCE). WCE represents new micro-sized robotic technology that uses a camera in a swallowable capsule to view and diagnose human's gastrointestinal (GI) tract. As the camera transmits sequences of color images and videos, it is a challenging task to effectively process the large amount of medical images and videos for diagnosis purpose. This chapter discusses relevant methods in video segments and event detection for the WCE application.

Power supply is of vital importance to the design of micro autonomous systems. Chapters 4 and 5 present new power supply techniques to replace traditional batteries. Specifically, Chap. 4 discusses the integration of nano/micro-fabrication practices and techniques to fuel cell system design. The proton exchange membrane fuel cell presents a comparative advantage over battery technology due to their higher energy density and ability to operate continuously without the need for recharging. It utilizes a polymer membrane, usually Nafion, to facilitate ion transport between the anode and cathode parts of the cell. Relevant MEMS technology for fuel cell construction is first presented, and then nano/micro-fabrication techniques for proton exchange membrane applications are discussed, together with experimental performance evaluation of Nafion membranes. It is desirable to use such a cheaper, more efficient, and better performance device to continuously power autonomous robots.

Chapter 5 presents vibration energy harvesting concept, methods, and potential utilization as a power source for micro/nano-robots. Energy harvesting refers to the process of converting excess energy in a given environment into a more useful form of electrical energy that can be used by the system. The chapter provides an excellent review of the general field of energy harvesting, after which mechanical vibration energy harvesting and other techniques with potentially greater application to biomedical nano/micro-robotics are discussed. In particular, different vibration energy harvesting mechanisms, such as electromagnetic, electrostatic, and piezoelectric techniques, are presented. The chapter concludes with recent work being done in the field of nanotechnology to further extend these energy harvesting approaches to size scales compatible with nano/micro devices and systems.

Chapter 6 of the book presents robot programming and control methods for a capsule robot navigating in the human's GI tract, which is conceptually designed to expand the functionality of the WCE discussed in Chap. 3. The chapter begins with a brief review of the available capsule endoscopy technology, which is followed by design principles and robot components of a conceptual design of the capsule endoscopic robot. Then the chapter focuses on navigation and control design, and introduces operating modes and navigation control of the robot. Behavior-based robot programming is introduced as the overall control architecture to organize sensor inputs, process sensing data, and coordinate actuation output effectively for real-time responding. To illustrate the navigation and control performance, robotic simulations are developed to visualize the navigation process of a capsule robot in the GI tract. Different behaviors, such as following the wall, collision avoidance, and object detection are demonstrated in the simulation. This chapter can serve as a case study in a course module to teach students in biomedical engineering the principles of robotics and programming.

Chapter 7 presents control methods for a multi-robotic system to cooperatively deliver drugs in tumor microenvironments. Based on a mathematical model of the ion diffusion process, the nanorobots can be pre-programmed to autonomously estimate the gradient of the center of the robot group using individual pH measurement, and then trace the estimated gradient to reach the tumor center with the lowest pH value. Convergence analysis of the control strategies is conducted using control theoretical methods, and numerical results are shown to illustrate the control performance in a simulated tumor microenvironment.

Chapter 8 discusses the challenges in the field of cancer targeted therapy, and focuses on targeted therapy at the individual cellular level using nanorobotic techniques. The rapid developments in nanorobots have allowed researchers to characterize the behavior and measure the interactions of single cells and single molecules, providing new insights into the mechanisms of cellular processes compared to the traditional ensemble experiments and bringing new techniques for discovering the novel biomarkers. The authors describe principles and methods of single molecule techniques using atomic force microscopy (AFM)-based nanorobotic manipulator, which include cell immobilization, mechanical property measurement, and measurement of protein–protein binding forces. The chapter provides practical knowledge on nanorobotic techniques to investigate cellular activities at the individual cell/molecule levels, which will have a great impact on disease treatment, drug development, and the development of rapid and high-efficiency detection method for molecular targeted therapy.

Chapter 9 presents a related topic on cell manipulation using optical tweezers, which function as special robotic end-effectors to impose force and deformation on a micro scaled object in a noninvasive manner. The chapter introduces the integration of robotics techniques into an optical tweezer system for single cell manipulation, where cell stretching and cell position are particularly discussed with experimental results. Through comparison between modeling and experiments, cell biomechanics property can be further characterized.

The last chapter of the book, Chap. 10, presents the design of a robot-assisted catheter surgery system and its remote operation though the Internet. Such a robotic system is expected to meet the challenges in intravascular neurosurgery. The system consists of a controller to be placed at the doctor side, and a catheter manipulator to be placed at the patient side, which are operated in a master–slave structure. This chapter describes the design of the master and slave machines, the communication between them, and a force feedback control to measure resisting force acting on the catheter, which is then sent to the controller to generate a haptic feedback to the surgeon at the master side. Real-time force monitoring methods are discussed, and real experiments are shown for performance evaluation of the system. An impressive two-way remote surgery experiment was carried out between Japan and China, which indicates the success of the two ways operation of the system.

Each of the above chapters is self-contained and can be used separately as a research or teaching reference. Besides the subject connections of the chapters described at the beginning of this subsection, the chapters can also be connected in either the techniques presented or applications used. For example, Chaps. 2, 4, and 5 have cross-referring *nanotechnology* components for advanced sensing and power generation. Chapters 3 and 6 are both on the *wireless capsule endoscopic* technology, with one focusing on biomedical imaging processing, and the other on robotic techniques for navigation. Chapters 8, 9 and 10 present *robotic manipulation techniques* including AFM-based robot manipulator, optical tweezers, and catheters, for applications in targeted cancer therapy, cell characteristics, and neurosurgery, respectively.

There is far more to cover in this multi-disciplinary field of micro/nanorobotics than can be contained in the single edited book. We hope the selected topics will inspire some further interest, and provide the reader necessary introduction and knowledge to pursue further study or investigation in the promising field.

References

1. Freitas RA Jr (2000) Say ah! in The New York Academy of Sciences, July/August 2000
2. Freitas RA Jr (1999) Basic capabilities. Nanomedicine, vol I. Landes Bioscience, Georgetown
3. Drexler KE (1992) Nanosystems: molecular machinery, manufacturing and computation. John Wiley and Sons, New Jersey
4. Tuszynski JA, Kurzynski M (2003) Introduction to molecular biophysics. CRC Press, Boca Raton
5. Rogers B, Pennathur S, Adams J (2008) Nanotechnology: understanding small systems. CRC Press, Boca Raton
6. Nelson BJ, Dong L, Arai F (2008) Micro/nanorobots. In: Siciliano B, Khatib O (eds) Springer handbook of robotics. Springer, Heidelberg, pp 411–450
7. Nelson BJ, Kaliakatsos IK, Abbott JJ (2010) Microrobots for minimally invasive medicine. Annu Rev Biomed Eng 12:55–85
8. Freitas RA Jr (2005) Current status of nanomedicine and medical nanorobotics. J Comput Theor Nanosci 2:1–25

9. Ummat A, Sharma G, Mavroidis C, Dubey A (2005) Bio-nanorobotics: state of the art and future challenges. In: Bronzino JD (ed) Tissue engineering and artificial organs, The biomedical engineering handbook. CRC Press, Boca Raton
10. Brehm M, Taubner T, Hillenbrand R, Keilmann F (2006) Infrared spectroscopic mapping of single nanoparticles and viruses at nanoscale resolution. Nano Lett 6(7):1307–1310
11. Rajangam K, Behanna HA, Hui MJ, Han X, Hulvat JF, Lomasney JW, Stupp SI (2006) Heparin binding nanostructures to promote growth of blood vessels. Nano Lett 6(9):2086–2090
12. Tozeren A, Byers SW (2004) New biology for engineers and computer scientists. Pearson Prentice Hall, New Jersey
13. Goedecke MD, Elston TC (2005) A model for the oscillatory motion of single dynein molecules. J Theo Biol 232:27–39
14. Berg H (2003) The rotary motor of bacterial flagella. Annu Rev Biochem 72:19–54
15. Blair DF (2003) Flagellar movement driven by proton translocation. FEBS Lett 545:86–95
16. Fischer S, Windshgel B, Horak D, Holmes KC, Smith JC (2005) From the cover: structural mechanism of the recovery stroke in the myosin molecular motor. Proc Nat Acad Sci 102:6873–6878
17. Jacobson SA, Epstein AH (2003) An informal survey of power MEMS, The international symposium on micro-mechanical engineering, Dec 1–3, 2003
18. Bullen RA, Arnot TC, Lakeman JB, Walsh FC (2006) Biofuel cells and their development. Biosens Bioelectron 21:2015–2045
19. Barton SC, Gallaway J, Atanassov P (2004) Enzymatic biofuel cells for implantable and microscale devices. Chem Rev 104:4867–4886
20. Xi J, Schmidt JJ, Montemagno CD (2005) Self-assembled microdevices driven by muscle. Nat Mater 4:180–184
21. DiSalvo FJ (1999) Thermoelectric cooling and power generation. Science 285 (5428):703–706
22. Collinsemail CM, Yang B, Yang QX, Smith MB (2002) Numerical calculations of the static magnetic field in three-dimensional multi-tissue models of the human head. Magn Reson Imaging 20(5):413–424
23. Siauve N, Scorretti R, Burais N, Nicolas L, Nicolas A (2003) Electromagnetic fields and human body: a new challenge for the electromagnetic field computation. Int J Comput Math Electr Electron Eng 22(3):457–469
24. Bath J, Turberfield AJ (2007) DNA nanomachines. Nat Nanotechnol 2:275–284
25. Guo Y, Zhang S, Man H, Ritter A (2011) A case study on pill-sized robot in gastro-intestinal tract to teach robot programming and navigation. Proceedings of ASEE annual conference and exposition, AC 2011-644, Vancouver, Canada, June 26–29, 2011
26. Alterovitz R, Goldberg K (2008) Motion planning in medicine: optimization and simulation algorithms for image-guided procedures, Springer, Heidelberg
27. Adriano C, Freitas R Jr (2005) Nanorobotics control design: a collective behavior approach for medicine. IEEE Trans Nanobiosci 4(2):133–140
28. Kim B, Park S, Jee C, Yoon S (2005) An earthworm-like locomotive mechanism for capsule endoscopes. Proceedings of IEEE/RSJ international conference on intelligent robots and systems, pp 2997–3002
29. Guo S, Pan Q (2006) Design and control of a novel type of microrobot moving in pipe. Proceedings of IEEE international conference on mechatronics and automation, pp 649–653, Luoyang, China, June 25–28, 2006
30. Selvakumar D, Chiang Y, Sims S, Wibbenmeyer J (2004) BioMEMS for future drug discovery needs. Technologies and services

Chapter 2
PZT Nano Active Fiber Composites-Based Acoustic Emission Sensor

Xi Chen and Yong Shi

Abstract A concept that utilizes the lead zirconate titanate (PZT) nano active fiber composites (NAFCs) to sense the acoustic emission (AE) for structural health monitoring (SHM) is demonstrated. The developed NAFCs consist of nanoscale PZT fibers with interdigitated electrodes on a silicon substrate. PZT nanofibers fabricated by an electrospinning process have a diameter of approximately 80 nm and aligned across the electrodes. Polydimethylsiloxane (PDMS) is severed as the polymer matrix and covers the sensor surfaces to protect nanofibers structures. The AE detection was demonstrated by both mounting the sensor on the surface of a steel table and embedding it in an epoxy structure. The output voltage reached the amplitude of 0.2 V in response to the acoustic wave generated by periodic impacts using a steel bar. The signal attenuation curves were measured to characterize the properties and demonstrate the anisotropic sensitivity of the PZT NAFCs sensor. The highly sensitive micro-sized PZT NAFCs sensors have promising applications in the SHM of composite structures such as the advanced carbon fiber-reinforced composites used in aerospace industry.

X. Chen · Y. Shi (✉)
Department of Mechanical Engineering, Stevens Institute of Technology,
Castle Point on Hudson, Hoboken, NJ 07030, USA
e-mail: Yong.Shi@stevens.edu

X. Chen
e-mail: xchen1@stevens.edu

Y. Guo (ed.), *Selected Topics in Micro/Nano-Robotics for Biomedical Applications*,
DOI: 10.1007/978-1-4419-8411-1_2, © Springer Science+Business Media New York 2013

2.1 Introduction

Recently, the application of carbon fiber composite structure has grown rapidly in the aerospace industry due to its high mechanical strength, high stiffness, and minimal weight [1, 2]. Being able to analyze and identify damages within composite structures using embedded sensors is highly desired [3]. Currently, strain gauges [4, 5], fiber optic sensors [6, 7], MEMS accelerometers [8], and piezoelectric acoustic emission (AE) sensors [9] are commonly used for structural health monitoring (SHM) applications. Piezoelectric sensors can achieve active sensing (excitation and sensing) [10] and the three other kinds of sensors are based on passive sensing. However, most of them are unsuitable for integrating into composites due to the small scale of the carbon fibers and the sensing technology. The active fiber composites (AFCs), developed first by Bent and Hagood [11, 12], is one of the great candidates for this application. A similar lead zirconate titanate (PZT) fiber composite, macro fiber composites (MFCs) developed by NASA [13], has also been used for such purposes. The MFCs with micron-scale PZT fibers embedded in a polymer matrix have demonstrated a number of advantages compared to other AE sensors [9]. Compared with the carbon fibers involved, the larger diameter and high stiffness of PZT fiber make it difficult to embed into composite structures. A potential solution to this issue is to replace the micro size PZT fiber with nanoscale fiber in MFCs.

Piezoelectric AE sensors are designed using a direct piezoelectric effect. The sensitivity of these kinds of sensors is mainly determined by the piezoelectric voltage constant (g_{33}) of the piezoelectric materials. The polyvinylidene fluoride (PVDF) thin film-based AE sensors have much less sensitivity than PZT sensors and cannot be used at a high temperature [14]. PZT nanofibers prepared by an electrospinning process exhibit an extremely high piezoelectric constant, high bending flexibility, and high mechanical strength, which have been demonstrated in our previous work [15]. The g_{33} of PZT nanofiber is roughly 0.079 Vm/N [15] which is much higher than that of the PZT bulk (0.025 Vm/N) or PZT microfiber (0.059 Vm/N) [16]. For a given energy input of the elastic waves, PZT nanofiber can produce much higher voltage than the PZT microfiber, which results in a higher sensitivity of the sensor. Unlike the bulk or micro size PZT ceramic materials, the high flexibility allows PZT nanofiber to bend freely in a transverse direction. Thus, the PZT nanofiber-based AE sensor can be easily bonded to curved structures. When implemented in a practical application, the PZT nanofiber has a smaller chance of being damaged due to its small size, flexibility, and high mechanical strength, which can extend the lifetime of nanofiber-based devices. The latest developments of carbon nanotube (CNT) composite-based strain sensors [17, 18] provide a way to achieve the SHM of composite structures. However, the passive sensing technique requires a current passing through the sensor, which generates heat and costs more energy. The PZT nanofiber-based nanogenerator that can covert mechanical energy into electrical energy has been successfully

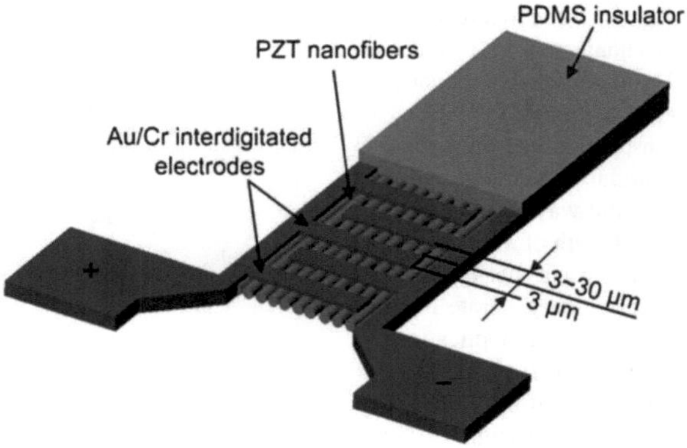

Fig. 2.1 Schematic view of the PZT NAFCs sensor. Reproduced from [1] by permission of the John Wiley and Sons (copyright 2011)

demonstrated in Ref [19] and illustrates the possibility of using the PZT nanofiber-based AE sensor to achieve the self-powered active sensing for SHM applications.

Here, we demonstrate an AE sensor that utilizes the PZT nano active fiber composites (NAFCs) for SHM. The developed PZT NAFC AE sensor consists of nanoscale PZT fibers with interdigitated electrodes on a silicon substrate. PZT nanofibers fabricated by an electrospinning process have a diameter varying from 60 to 85 nm and are aligned across the electrodes. Polydimethylsiloxane (PDMS) is severed as the polymer matrix and cover the sensor surfaces to protect nanofibers structures. The AE detection was demonstrated by both mounting the sensor on the surface of a steel table and embedding it in an epoxy structure. The output voltage reached the amplitude of 0.2 V in response to the acoustic wave generated by the periodic impacts using a steel bar. The signal attenuation curves were measured to characterize the properties and demonstrate the anisotropic sensitivity of the PZT NAFCs AE sensor. The highly sensitive micro-sized PZT NAFCs AE sensors have promising applications in the SHM of composite structures such as the advanced carbon fiber-reinforced composites used in aerospace industry.

2.2 Design of PZT NAFCs AE Sensor

2.2.1 PZT NAFCs AE Sensor Concept

The concept of PZT NAFCs AE sensor is presented in Fig. 2.1. The PZT NAFCs sensor consists of nanofibers, interdigitated electrodes, and a polymer matrix. The PZT nanofibers are first aligned on a silicon substrate. The gold interdigitated electrodes are directly evaporated onto the surface of the PZT nanofibers, while the polymer layer is coated on top of the electrodes for mechanical integrity. The width of the deposited electrode was 3 μm and the gap between the two adjacent electrodes varied from 3 to 20 μm. The thickness of the whole sensor can be controlled within 2 μm when the PZT NAFCs AE sensor is released from the silicon substrate.

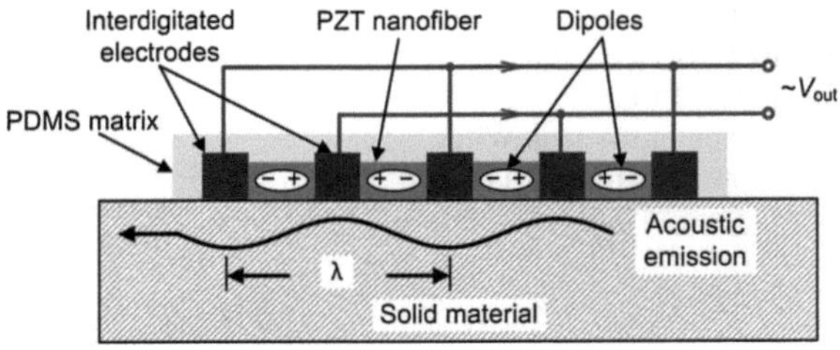

Fig. 2.2 The AE detection mechanism of PZT NAFCs. The figure shows the side view of the PZT NAFCs sensor mounted on a solid material to be monitored. The acoustic wave with a wavelength of λ illustrates the typical wave that PZT NAFCs can generate for the active sensing application. Reproduced from [1] by permission of the John Wiley and Sons (copyright 2011)

2.2.2 AE Detection Mechanism

The AE detecting mechanism of PZT NAFCs is illustrated in Fig. 2.2. The acoustic waves transferred through the structures, which induced elastic stress waves along the aligned nanofibers. The charge was separated in those nanofibers by the longitudinal stress waves, which resulted in a piezoelectric potential gradient along the nanofiber [20]. The generated piezoelectric potential $\sim V_{\text{out}}$ was picked up through the adjacent electrodes and transported to an external detection circuit.

The voltage generated between two adjacent electrodes of the PZT NAFCs can be calculated by

$$V_{\text{out}}^{1} = \int g_{33}\varepsilon_{33}(l)E_{\text{p}}\text{dl} \tag{2.1}$$

where $\varepsilon_{33}(l)$ is the longitudinal strain along PZT nanofibers, l is the length of the nanofibers across the two adjacent electrodes, and E_{p} is the Young's modulus of PZT nanofibers. Assuming a perfect bonding between PZT nanofibers and PDMS polymer matrix, we have

$$\varepsilon_{\text{p}} = \varepsilon_{\text{m}} = \varepsilon_{\text{c}} \tag{2.2}$$

where ε_{p} is the strain in PZT nanofibers, ε_{m} and ε_{c} are the strains in the PDMS matrix and composite. Assuming both PZT nanofibers and PDMS matrix is elastic, the PZT NAFCs can be considered as unidirectional continuous fiber lamina. Therefore, we have

$$E_{33} = E_{\text{p}}\text{v}_{\text{p}} + E_{\text{m}}(1 - \text{v}_{\text{p}}) \tag{2.3}$$

$$\text{v}_{\text{p}} = A_{\text{p}}/A_{\text{c}} \tag{2.4}$$

where E_{33} is the longitudinal modulus, E_m is the Young's modulus of PDMS polymer, A_p is net cross-sectional area for the fibers, and A_c is the cross-sectional area for the PZT NAFCs. The major Poisson's ratio could be calculated as

$$\upsilon_{31} = \upsilon_{32} = \upsilon_p v_p + \upsilon_m(1 - v_p) \tag{2.5}$$

where υ_p and υ_m are the Poisson's ratio of PZT nanofibers and PDMS polymer, respectively. The longitudinal strain in the PZT nanofibers can be given as

$$\varepsilon_{33}(l) = \sigma_{33}(l)/E_{33} - \sigma_{11}(l)\upsilon_{31}/E_{33} - \sigma_{22}(l)\upsilon_{32}/E_{33} \tag{2.6}$$

where $\sigma_{11}(l)$, $\sigma_{22}(l)$ and $\sigma_{33}(l)$ are the stress functions in three directions. Thus, the output voltage can be rewritten as

$$V_{out}^1 = \int \left[\frac{\sigma_{33}(l)}{E_{33}} - \frac{\sigma_{11}(l)\upsilon_{31}}{E_{33}} - \frac{\sigma_{22}(l)\upsilon_{32}}{E_{33}}\right] E_p g_{33} \mathrm{dl} \tag{2.7}$$

According to Eq. (2.7), the output voltage from the PZT NAFCs sensor is a function of the strength of the elastic wave. Since the PZT nanofibers were polled in opposite directions between each adjacent electrode (see Fig. 2.2), the output power was enhanced by the parallel connection of those units using an interdigitated pattern of electrodes. For the application of active sensing, the PZT NAFCs can generate elastic waves with the wavelength of λ, which is equal to twice the distance [21] between the electrodes as shown in Fig. 2.2.

2.3 Fabrication

2.3.1 PZT Nanofiber Synthesization

PZT nanofibers were prepared by electrospinning process [22, 23]. The starting material, PZT (52/48) sol–gel and poly vinyl pyrrolidone (PVP, Aldrich) were obtained from commercial sources. Alcohol was used as a solvent for PVP, while acetic acid was added to stabilize the solution and to control the hydrolysis reaction of the sol–gel precursor [24]. After being stirred vigorously for 2 h at room temperature, the mixture was fed into a microscale metallic tube through a syringe pump. A droplet of the modified sol–gel solution was held at the orifice of the metallic tube due to the high surface tension. A high DC voltage (10 kV) between the tip of the micro metallic tube and the collecting substrate was applied to overcome the surface tension and generate an electrically charged jet of the modified solution. After electrospinning, the as-spun nanofibers were annealed at 650 °C to obtain PZT nanofibers with the perovskite phase.

Figure 2.3 has shown the morphology and X-ray diffraction pattern of the annealed PZT nanofibers. Figure 2.3a was the scanning electron microscopy (SEM) image of a selected nanofiber with a diameter of 70 nm. By varying the

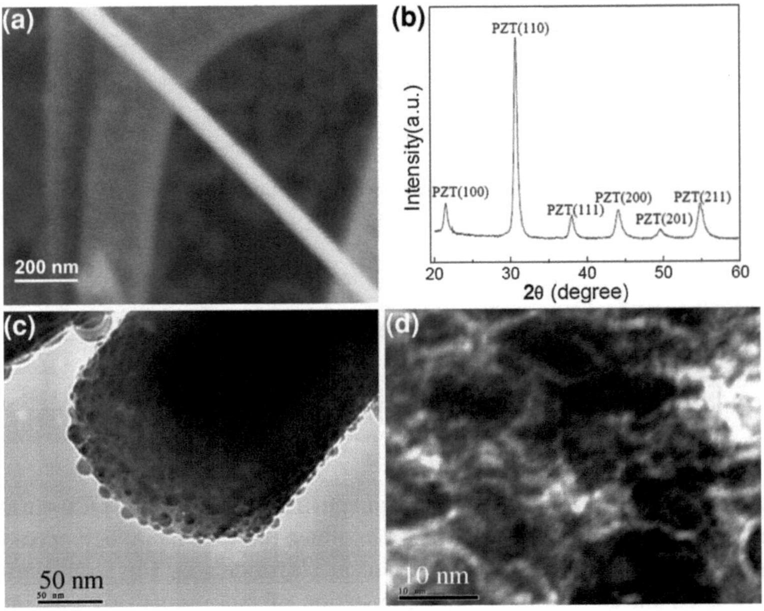

Fig. 2.3 **a** SEM image of single PZT nanofiber. **b** X-Ray diffraction pattern for the PZT nanofibers. **c** and **d** TEM images of a PZT nanofiber. Reproduced from [15] by permission of the American Institute of Physics (copyright 2009)

concentration of the polymer in the precursor, the average diameters of the nanofibers could be tuned from 52 ± 8 nm to 150 ± 12 nm. The X-ray diffraction (XRD) pattern of the annealed PZT nanofibers shown in Fig. 2.3b indicated that the PZT nanofibers annealed at 650 °C had pure perovskite phase. The annealing temperature was lower than that used by most other methods. Transmission electron microscopy (TEM) images of the cross-section and surface of PZT nanofibers were shown in Fig. 2.3c and d. The grain size of the PZT nanofibers was about 10 nm. The structure of the nanofiber seemed porous probably because of the polymer precursor solution of the sol–gel process [25]. By modifying the precursor and controlling the electrospinning parameters, a denser nanostructure could be obtained.

2.3.2 Fabrication of PZT NAFCs AE Sensor

The designed fabrication process is as shown in Fig. 2.4. A silicon wafer was used as the substrate to collect nanofibers and also to fabricate the composites (step a.). Then, a thin layer of thermal oxide was grown on the surface of the substrate as a diffusion barrier to prevent silicon transfer into the piezoelectric (PZT) fibers at elevated temperature and change their composition (step b.). Aligned piezoelectric

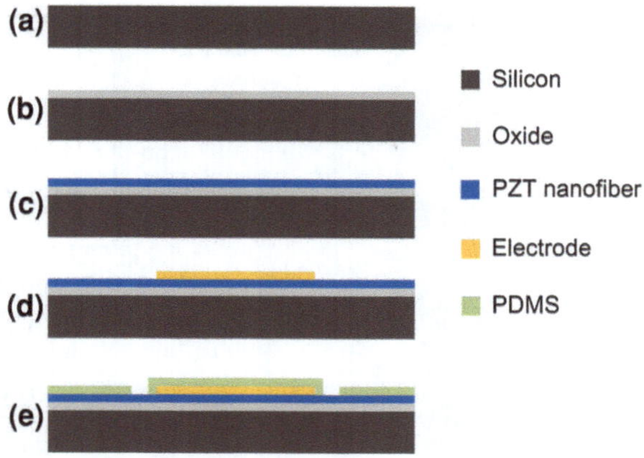

Fig. 2.4 Fabrication process of PZT NAFC AE sensor

nanofibers were deposited onto the substrate by electrospinning followed by an annealing process at 650 °C for 25 min (step c). A lift-off process was used to prepare interdigitated electrodes by using the photoresist of AZ5214 (step d). Polymer (PDMS) layer was deposited by spin coating (step e). Finally, the PZT nanofibers were polled by applying an electric field of 4 V/μm across the interdigitated electrodes at a temperature of 140 °C for about 24 h.

As shown in Fig. 2.5, a millimeter size PZT NAFCs AE sensor was fabricated on a silicon substrate. After the annealing process, PZT nanofibers with a diameter of approximately 80 nm (see Fig. 2.5a) had a large concentration of perovskite phase as shown in the XRD pattern (inset in Fig. 2.5a). As shown in Fig. 2.5b, the PZT nanofibers were entirely covered by the thermally evaporated Au/Cr interdigitated electrodes to achieve a good connection. The width of the deposited electrode was 3 μm and the gap between the two adjacent electrodes varied from 3 to 20 μm.

By varying the size of interdigitated electrodes, the size of a single PZT NAFCs sensor cell can be fabricated from millimeter scale to microscale. Figure 2.6 shows a microscale sensor that has been fabricated.

After applying a thin PDMS layer as the matrix material on the top of the sensor, the wafer was cut to form the individual AE sensors. The finished PZT NAFCs sensor cells are as shown in Fig. 2.7. The silicon substrate was used as the rigid mechanical backing to avoid excessive stresses on PZT nanofibers and minimize the risk of damaging the electrical connection. The extraction copper wires were connected with the electrodes by using silver glue to connect with the external circuits. The PZT NAFCs AE sensor can be released from the silicon substrate by dry etching for different applications.

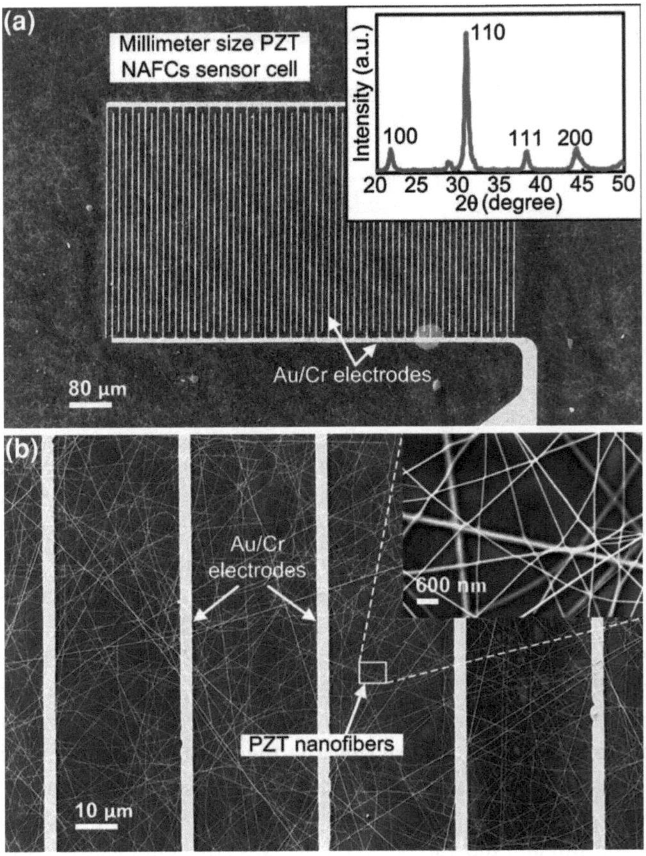

Fig. 2.5 **a** A SEM image of the millimeter size PZT NAFCs sensor cell before coating the PDMS layer. The inset in **a** shows the XRD pattern for the PZT nanofibers. **b** A SEM image of PZT nanofibers covered by Au/Cr interdigitated electrodes. The inset in **b** shows the diameter of PZT nanofiber is approximately 80 nm. Reproduced from [1] by permission of the John Wiley and Sons (copyright 2011)

Fig. 2.6 A SEM image of a micron size PZT NAFCs sensor cell. Reproduced from [1] by permission of the John Wiley and Sons (copyright 2011)

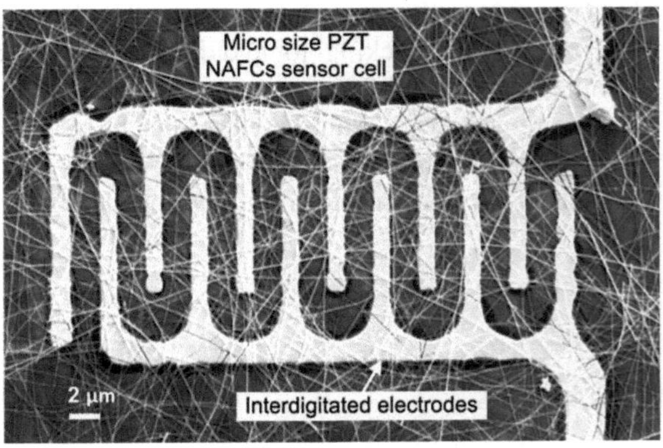

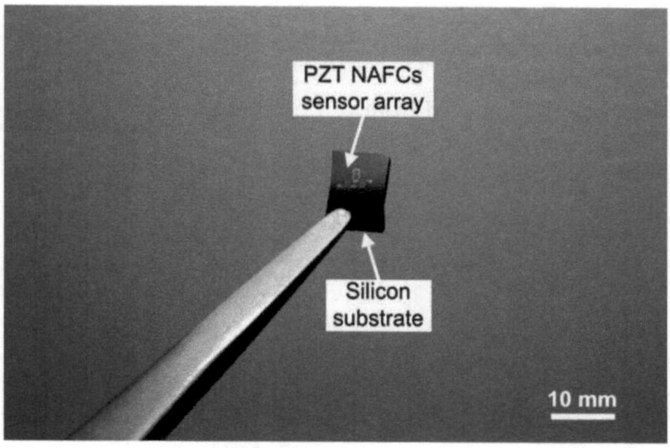

Fig. 2.7 An optical image of PZT NAFCs sensor cell array after coating a thin PDMS layer on the top surface. Reproduced from [1] by permission of the John Wiley and Sons (copyright 2011)

2.4 Characterization of PZT NAFCs AE Sensor

2.4.1 Elastic Wave Detection

A single AE sensor cell, with the width of 400 μm and length of 800 μm, was selected to perform the measurement. The experiment was first carried out for the surface elastic wave detection. The PZT NAFCs was reversibly mounted on a grounded steel surface as shown in upper inset in Fig. 2.8. To eliminate the influence of the electromagnetic interference, the sensor structure was fully covered using a faraday cage. A grounded steel bar was used to generated the acoustic wave along the parallel direction of the nanofibers by periodic knocking on the steel surface. As shown in Fig. 2.8, the charge separation was recorded when the elastic wave reached to the NAFCs. The sensed signal was induced by piezopotential-driven transient flow of electrons under the stress of the substrate generated by the waves [26]. The higher the impact energy applied on the steel surface, the higher the voltage generated from the PZT NAFCs. The measured peaks were around 0.08 V during the test and the signal vanished fast due to the low energy.

The PZT nanofiber composites was then embedded into a square-shaped epoxy structure and fine copper wires were utilized as the extraction electrodes to transport electrons to external circuit, see upper inset in Fig. 2.9. As shown in Fig. 2.9, the positive and negative voltage outputs were observed when a grounded steel bar was knocking on the epoxy structure and the voltages reached to 0.2 V during the test. The negative voltage distribution was generated due to the reversely flowing carriers when the imbedded NAFCs suffering the restoring force from the structure oscillation after the external load was removed. Therefore, the PZT NAFCs AE sensor shows a promising application in SHM by integrating into structures.

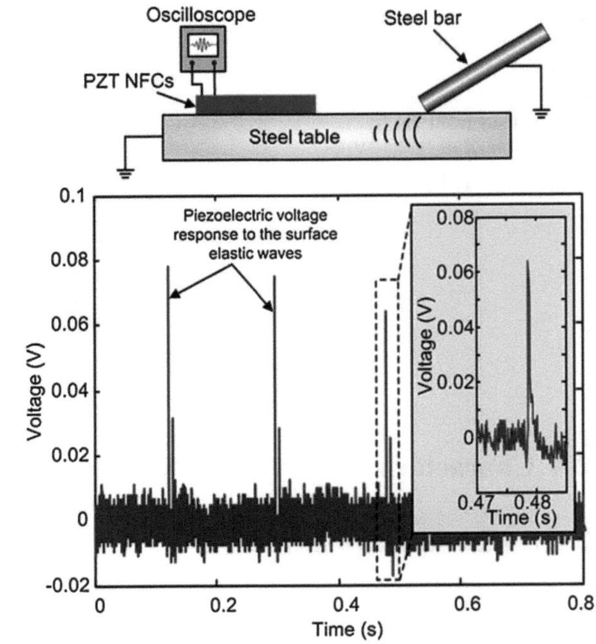

Fig. 2.8 The voltage generation when used for surface elastic wave detection. Reproduced from [1] by permission of the John Wiley and Sons (copyright 2011)

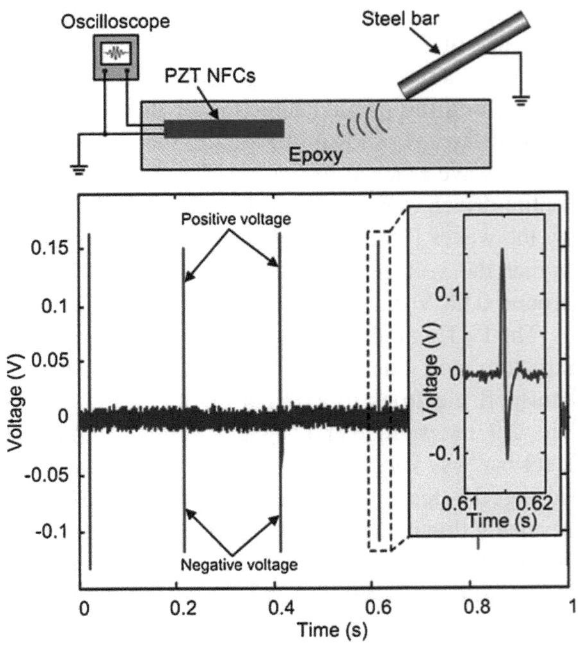

Fig. 2.9 The voltage generation when the PZT NAFCs was embedded in an epoxy structure. Reproduced from [1] by permission of the John Wiley and Sons (copyright 2011)

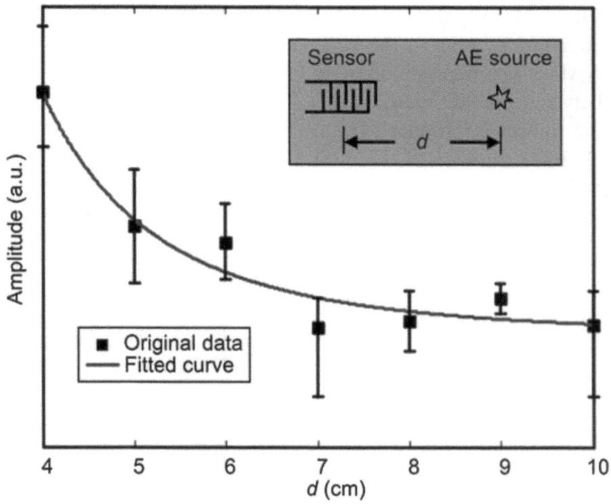

Fig. 2.10 Distance attenuation curve of the PZT NAFCs sensor as a function of distance between the AE source and the sensor. Reproduced from [1] by permission of the John Wiley and Sons (copyright 2011)

2.4.2 Signal Attenuation

In order to further study the sensing properties of the NAFCs sensor, the signal attenuation curves were measured by changing the location of the AE source on an isotropic Poly(methyl methacrylate) (PMMA) test plate. The signal attenuation test was first performed as a fucntion of distance (d) along a direction parallel to the PZT NAFCs, as seen in the inset in Fig. 2.10. The PZT NAFCs were permanently bonded on the top surface of the PMMA test plate using epoxy. The AE signal was generated by utilizing the impact energy of a metal ball dropped on the test plate from a certain height. By varying the distance of AE source from 4 to 10 cm, the average amplitude of the output voltage was recorded at each location after several tests as shown in Fig. 2.10. From the measurement, it was determined that the values also depend on the shape of the PMMA plate and the sensor locations.

According to Eq. (2.7), the piezoelectric potential only indicates the longitudinal deformation of PZT nanofibers. The elastic waves along the nanofibers direction produce higher voltage than the waves normal to the fiber direction. The measurement of angular attenuation was then carried out by fixing the distance at 5 cm and changing the angles between the AE source and the sensor orientation. By varying the polar angle, the angular attenuation curve was measured as shown in Fig. 2.11. Therefore, the aligned nanofibers and interdigitated electrodes can lead to the anisotropic sensitivity of PZT NAFCs sensors. With the distance and angular attenuation properties, the AE source information can be easily identified using a small number of sensors.

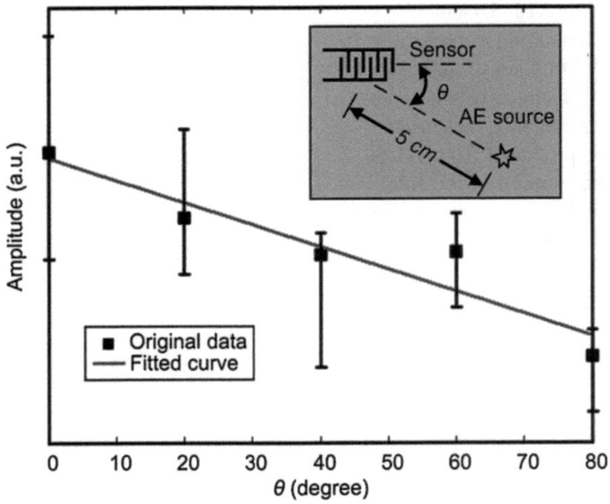

Fig. 2.11 Angular attenuation curve of the PZT NAFCs as a function of angle between the AE source and the sensor. Reproduced from [1] by permission of the John Wiley and Sons (copyright 2011)

2.4.3 AE Detections of Micron Size Sensor

The AE detections of micron size PZT NAFCs sensors were also demonstrated by permanently mounting on the top surface of PMMA. For less PZT nanofibers and interdigitated electrodes, the sensitivity of micron size PZT NAFCs sensor is much lower than that of millimeter size sensor. SR560 preamlifier was used to amplify the generated voltage during the AE detection. A band-pass RC filter with cutoff frequencies of 300 Hz and 1 MHz was used to reduce the environmental noise to enhance the sensitivity of the sensor. Two micron size PZT NAFCs sensors, with the width of 15 μm and length of 30 μm, were used to sense the AE signal as shown in Fig. 2.12. With the source of periodic knocking on the PMMA surface using a steel bar, the amplitude of those sensors was about 10 mV which is much lower than that of the larger sensors described in the earlier section. The sensitivity can be enlarged by precisely designing the gaps between electrodes for different materials to be monitored [27].

During the test, all equipment and structures were well grounded in order to eliminate the influence of the bioelectric field of the human body and the electromagnetic interference from the testing equipment. The faraday cage was involved and well grounded during experiment. The amplitude of the noise signal was controlled under 2 mV. An interdigitated electrode based silicon substrate without PZT nanofibers had also been tested to verify the piezoelectric phenomena of NAFCs when subjected to the elastic waves. Although the surface charge of the devices could also create the potential difference, those signals can be eliminated by correct grounding.

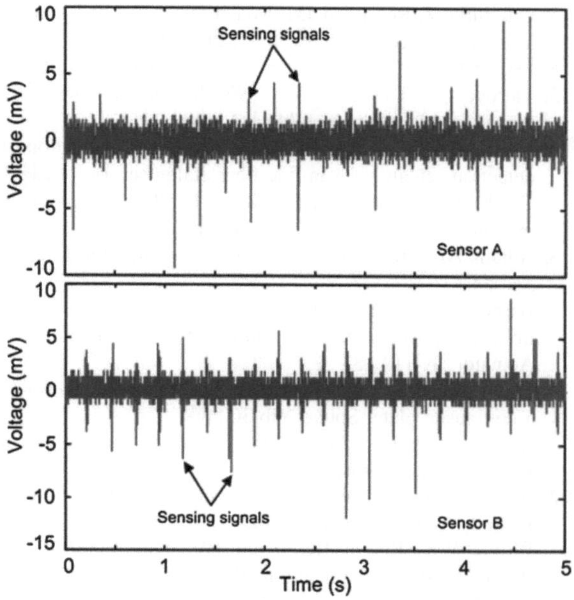

Fig. 2.12 The demonstration of micron size PZT NAFCs used for acoustic emission detection. **a** The voltage generation when used for AE detection of sensor A. **b** The voltage generation when used for AE detection of sensor B. Reproduced from [1] by permission of the John Wiley and Sons (copyright 2011)

2.5 Conclusion

In conclusion, a new concept of using PZT NAFCs with interdigitated electrodes as AE sensors for SHM application was demonstrated. The diameter of PZT nanofibers was controlled at ~80 nm. The acoustic responses of the PZT NAFCs sensor were demonstrated by mounting the sensor on the surface of the steel table and embedding it in the epoxy structure. The peak voltage output reached 0.2 V in response to the acoustic wave generated by the periodic impacts using a grounded steel bar. The signal attenuation curves showed the anisotropic sensitivity of the PZT NAFCs sensor. The high sensitivity, microscale size, low weight, large flexibility, low cost, and anisotropic sensitivity make the PZT NAFCs sensors have a promising application in health monitoring by integrating them into composites. The active PZT materials also allow the possibility of the NAFCs sensor to achieve self-powered active sensing for the SHM application.

Acknowledgments This work was supported in part by the National Science Foundation (Award No. CMMI-0826418 & No. ECCS-0802168). The authors would also like to thank J. Li and G. Zhang for useful discussions.

References

1. Chen X, Li J, Zhang G et al (2011) PZT nanoactive fiber composites for acoustic emission detection. Adv Mater 23:3964
2. Lin M, Chang FK (2002) The manufacture of composite structures with a built-in network of piezoceramics. Compos Sci Technol 62:919

3. Giugiutiu V, Redmond JM, Roach DP et al (2000) Active sensors for health monitoring of aging aerospace structures. SPIE Proc 3985:294
4. Perry CC (1987) Strain gauge measurements on plastics and composites. Strain 23:155–156
5. Tairova LP, Tsvetlov SV (1993) Specific features of strain measurement in composite materials. Mech Compos Mater 28:489–493
6. Singh H, Sirkis JS, Andrews J et al (1995) Evaluation of integrated optic modulator-based detection schemes for in-line fiber etalon sensors. J Lightwave Technol 13:1772
7. Todd MD, Johnson GA, Vohra ST (2001) Deployment of a fiber bragg grating-based measurement system in a structural health monitoring application. Smart Mater Struct 10:534
8. Adams DE (2007) Health monitoring of structural materials and components. Wiley, West Sussex
9. Barbezat M, Brunner AJ, Flueler P et al (2004) Acoustic emission sensor properties of active fibre composite elements compared with commercial acoustic emission sensors. Sens Actuators A 114:13
10. Park HW, Sohn H, Law KH et al (2007) Time reversal active sensing for health monitoring of a composite plate. Sound Vib 302:50
11. Bent A, Hagood NW (1995) Anisotropic actuation with piezoelectric fiber composites. J Intell Mater Syst Struct 3:338
12. Bent A, Hagood NW (1997) Piezoelectric fiber composites with interdigitated electrodes. J Intell Mater Syst Struct 8:903
13. Wilkie WK, Bryant RG, High JW et al (2000) Low-cost piezocomposite actuator for structural control applications. Proc SPIE 3991:323
14. Monkhouse RSC, Wilcox PW, Lowe MJS et al (2000) The rapid monitoring of structures using interdigital lamb wave transducers. Smart Mater Struct 9:304
15. Chen X, Xu SY, Yao N et al (2009) Potential measurement from a single lead ziroconate titanate nanofiber using a nanomanipulator. Appl Phys Lett 94:253113
16. Swallow LM, Luo JK, Siores E et al (2008) A piezoelectric fibre composite based energy harvesting device for potential wearable applications. Smart Mater Struct 17:025017
17. Thostenson ET, Chou TW (2006) Carbon nanotube networks: sensing of distributed strain and damage for life prediction and self healing. Adv Mater 18:2837
18. Yamada T, Hayamizu Y, Yamamoto Y et al (2011) A stretchable carbon nanotube strain sensor for human-motion detection. Nat Nanotechnol 6:296
19. Chen X, Xu SY, Yao N et al (2010) 1.6 V nanogenerator for mechanical energy harvesting using PZT nanofibers. Nano Lett 10:2133
20. Xu S, Qing Y, Xu C et al (2010) Self-powered nanowire devices. Nat Nanotechnol 5:366
21. Monkhouse RSC, Wilcox PD, Cawley P (1997) Flexible interdigital PVDF transducers for the generation of Lamb waves in structures. Ultrasonics 35:489
22. Chen X, Yao N, Yong S (2011) Energy harvesting based on PZT nanofibers. In: Zhang L (ed) Energy efficiency and renewable energy through nanotechnology. Springer Series in Nanoscience and Technology
23. Xu SY, Shi Y, Kim SG (2006) Fabrication and mechanical property of nano piezoelectric fibers. Nanotechnology 17:4497–4501
24. Li D (2003) Xia Y (2003) Fabrication of tatania nanofibers by electrospinning. Nano Lett 3:555–560
25. Li D, Wang Y, Xia Y (2004) Electrospinning nanofibers as uniaxially aligned arrays and layer-by-layer stacked films. Adv Mater 16:361–366
26. Wang ZL (2010) Piezotronic and piezophototronic effects. J Phys Chem Lett 1:1388
27. Zhu W, Rose JL (1999) Lamb wave generation and reception with time-delay periodic linear arrays: a BEM simulation and experimental study. IEEE Trans Ultrason 46:654–664

Chapter 3
Biomedical Image Analysis on Wireless Capsule Endoscopy Images and Videos

Guang Yang, Yafeng Yin and Hong Man

Abstract Wireless capsule endoscopy (WCE) is a newly booming technology on gastrointestinal (GI) examinations. After the patient swallows the capsule, it starts taking video of the entire interior of the digestive tract, from the esophagus, stomach to small intestine, and colon. It transmits sequences of color images back to the computer, and hence allows the doctors to observe the patient's GI tract vividly and hence diagnose precisely based on the visually accessed video instead of reconstructed images such as MRI. The entire process takes hours with around 50,000 images about the inside of the GI tract. It is very time-consuming to go through those images frame by frame. Therefore, many image/video processing techniques are adopted to develop software which is expected to annotate the images in different stage of the GI tract and detect unusual events, such as bleeding, polyps, and ulcers, automatically. In another words, the problems in WCE images/videos can be categorized in two classes, video segmentations and event detections. Various methods have been discussed in the literatures and generally can be seen as two-stage approaches, describing images and classifying images or regions. In this chapter, we will review various feature descriptors and classification methods that are often used on WCE images/videos.

G. Yang (✉) · Y. Yin · H. Man
Department of Electrical and Computer Engineering, Stevens Institute of Technology, Hoboken, NJ 07030, USA
e-mail: gyang1@stevens.edu

H. Man
e-mail: Hong.Man@stevens.edu

Y. Guo (ed.), *Selected Topics in Micro/Nano-Robotics for Biomedical Applications*, DOI: 10.1007/978-1-4419-8411-1_3,

3.1 Introduction

Traditional endoscopy consists of a flexible fiber with a camera on the top which is pushed into the patients digestive tract, i.e., stomach, upper small bowel, or colon, and allows the doctor to visualize the inside surface of the digestive tract. This procedure has hardly a substitution; however, although the patients suffer a great discomfort during the invasive investigations, the traditional endoscopy is still unable to thoroughly check the entire gastrointestinal (GI) tract. Especially for the coiled small bowel, the farther the endoscopy gets the heavier pain the patient suffers. Also, the coiled shape limits the depth that the endoscopy could reach. The emerging new type of endoscopy [1, 2], which is wireless and in swallowable capsule size, has illumination sources and CMOS chip camera to take videos inside the tract as well as a radio transmitter and an antenna to transmit the video back to the terminal receiver.

During the inspection of the WCE video taken from the patients, doctors could easily distinguish the abnormal cases frame by frame. However, as the same reason of the critical length of the video going through the entire digestive tract, for example, an average length of a video is about 50,000 frames in the frame rate of 2 fps. Even if with the preprocessing that converts to a frame rate of 25 fps, the video still takes half an hour long. It is a very time-consuming task to review the video frame by frame, not to mention that there might be many more patients' videos are waiting in line. Computer algorithms emerged aiming at automatically detecting the changing of the stages or the unusual events in the GI tract, mark the locations, and hence help the doctors out of this dilemma.

Accordingly, the automatic detections in WCE images/videos can be categorized in two classes, video segmentation and events detection, as illustrated in Fig. 3.1. The video segmentation is to annotate meaningful stages of the GI tract [3, 4]. Namely, the mouth and esophagus as the entrance stage of swallowing the capsule, the capsule is arriving at the stomach and approximately staying for 80 min [1], and consequently passing through the pylorus and taking around 90 min [1] in the small intestine stage, and finally the colon is the last stage. And it is more important to segment the part of the small intestine stage, for the reason that there is hardly any other kind of endoscopy that could manage to access this part.

Doctors would diagnose the patients based on the videos, if there are any pathological lesions that have been viewed from the endoscopy images. Therefore, the further task is to automatically detect when events happen in the small intestine, especially for those pathology abnormalities [5, 6] such as bleeding, ulcer, polyp, and Crohn's disease. There are also non-pathological disturbance [5], i.e., juices and bubbles [7], which also need to be distinguished from the normal intestine wall and, particularly, the aforementioned abnormal patterns.

A lot of research has been done to achieve the tasks of video segmentation and abnormal event detection since the capsule endoscopy was approved by the U.S. Food and Drug Administration (FDA) in 2001 [8]. Various methods have been

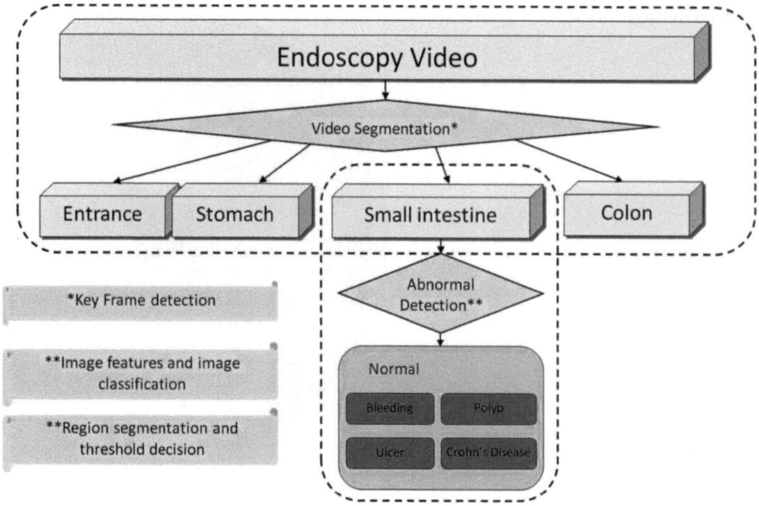

Fig. 3.1 Video segmentations and abnormal detections in small bowel are the two main problems in videos of the WCE

studied and generally can be seen as two-stage approaches, describing images, and classifying images or regions. To describe images, there are many sophisticated methods on feature extractions. Based on the literatures, many approaches of extracting color and texture features are used for recognizing the pattern of images. Meanwhile, there are various off-the-shelf machine learning methods that are capable to classify WCE images. In this chapter, we will review various feature descriptors and classification methods that often adopted on WCE images/videos.

3.2 Color Features

Unlike other videos taken from the daily lives, videos from WCE have relatively simple scenario. Therefore, colors can provide rich information of the current scene. Capsule endoscopy images are originally obtained in RGB color space, but most of the time, color images would be transformed into HSV/HSI color space for its good property of color invariant. In [9, 10], features are extracted in RGB color space. In other literatures [11–13], images are converted into HSV/HSI space before extracting features. In the study [14], statistical features are obtained from multiple color spaces at the same time.

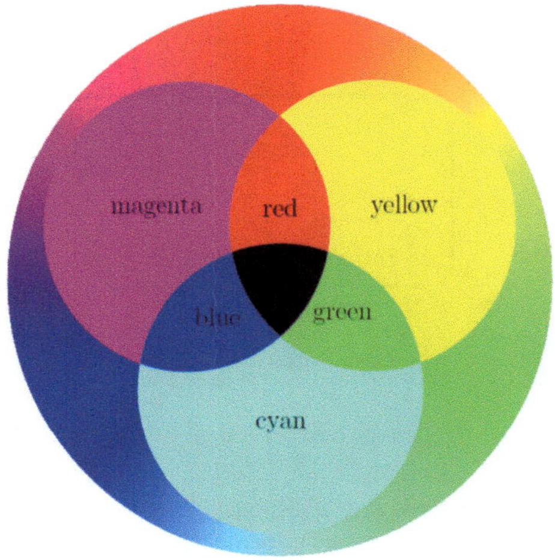

Fig. 3.2 RGB color space

3.2.1 Color Spaces and Transformations

In the field of image processing, RGB color space is one of the most basic and frequently used models as describing color images. As an additive color model, this color standard was initiated by CIE (the International Commission on Illumination) in 1931 designing each primary color with a specific wavelength as: $R = 700\,\text{nm}$, $G = 546.1\,\text{nm}$ and $B = 435.8\,\text{nm}$. The reason of choosing these three colors was not because of physics but the physiology of human visual system, in which photoreceptor cell, also known as opsin, senses the wavelength of lights, absorbs photons, and transmits the signal to hyperpolarize the photoreceptor. There are two major kinds of photoreceptor cells, rod opsin, and cone opsin. Rods can sense in weak light condition. And cones have three types, for each absorbing long, middle, and short wavelength of light corresponding to color red, green, and blue.

In Fig. 3.2, red and green together generate yellow, red added blue is magenta, and green with blue is cyan. These three colors are also known as secondary colors. It is clear that, primary colors can be obtained by subtractions of secondary colors. Therefore, while RGB are additive primary colors, yellow, magenta, and cyan are called subtractive primaries. Furthermore, although these color component are fixed, any color along all visible spectrum, can be generated by a mixture of these three components with different intensities and wavelengths.

Hue and saturation together are carrying color information (chromaticity), which best construct the human visual perception in the manner of measurement and interpretation. Consider a color is a mixture of light waves; hue is characterizing the principle wavelength band in it. Hue is the result of our eyes sensation of the color, and tells the color as we name them, green, orange, or red. Saturation refers to the proportion of the hue mixed with the white light, which is the mixture of full spectrum of visible light. The value, intensity, and lightness are slightly different on the definitions, which are listed in Eqs. (3.4), (3.5), and (3.6). The transformation from RGB space to Hue and Saturation space is shown as follows:

$$H = \begin{cases} \theta & \text{if} \quad G \geq B \\ 360 - \theta & \text{if} \quad G < B \end{cases} \tag{3.1}$$

$$\theta = \cos^{-1}\left\{\frac{R - \frac{1}{2}(G+B)}{\left[(R-G)^2 + (R-B)(G-B)^{\frac{1}{2}}\right]}\right\} \tag{3.2}$$

$$S = 1 - \frac{3}{R+G+B} \times \min(R, G, B) \tag{3.3}$$

$$I = \frac{1}{3}(R+G+B) \tag{3.4}$$

$$V = \max\{R, G, B\} \tag{3.5}$$

$$L = \frac{1}{2}\max\{R, G, B\} + \frac{1}{2}\min\{R, G, B\} \tag{3.6}$$

3.2.2 Color Features

Statistical features of each color channel in both RGB and HSV color space are taken by Kodogiannis and Boulougoura in [14] to estimate the histograms of the transformed image intensities. RGB color model of the gastrointestinal tract images are build to analyze the dominant color component in the normal GI tract regions [10]. A signal of the dominant color component of each frame is hence traced through the entire GI tract and about 80 % of the sequences correctly detected the changes between two adjacent organs.

HSI color space is more popularly used for the reason that hue itself gives the color as we described, as well as the discrimination property [4, 15] is more advanced than other color space. Many literatures have converted color image from RGB to HSI color space to prepare for the subsequent procedures. Simple segmentation is preformed in HSI color space in [12]. In this procedure, the information from hue channel is kept where its corresponding saturation is above a preset threshold. The complement of the intensity channel is also limited by the binary image and hence joint with the information from hue channel to finally obtain a segmented image from the color input. Coimbra and Cunha [16] have tested that the uniformly quantized color histograms in HSV color space could provide the most sufficient classification information on abnormal event such as blood, ulcer, or polyps.

In [3], two-dimensional histograms are built for hue and saturation channels which are modeling the HS distributions on different parts of the entire digestive tract, from the entrance of the mouth, stomach to small intestine and colon. In a more recent work [13], features obtained from RGB and HSI color space are combined to have even more robust candidates of the abnormal regions. Segmentations are preformed on RGB color image, where only the hue channel offers possible candidate region after Log-Gabor filters.

3.2.3 Histogram

Histogram of an image is the statistical distribution of the intensity values or the intensity ranges (bins) composing the image. A histogram of an image is a mapping function that gives the relationship between an amplitude value and the number of occurrence of that value. The probability of an amplitude value $p(r_i)$can be considered as the number of its occurrence $n(r_i)$over the total number of pixels.

One-dimension histogram is mostly used on the gray-scale image, for its single channel of intensity. Histograms can be the distribution of every value on full intensity brand, say [0, 255]. But to be more efficient, for those values close enough, they can be counted in one bin and hence the full intensity range is segmented by a set of bins evenly. For example, for a histogram with four bins, the values in the range of [0, 63] will be counted in the first bin, and similarly, the second bin is ranging in [64, 127], and so on.

Color histograms in two dimensions or three dimensions are often used for color images in RGB or HSV spaces, for instance. The color histogram is drawing a value in the image in a position of two or three axes set and counting the number of the values, same as in a one-dimensional histogram. In the color histogram, axes are divided in a set of bins, and a multi-channel color pixel will be found in a cross position by the bins in each axis.

Features in histogram domain are easy to obtain, and sometimes bring much information to classify. The histogram of intensity reveals the contrast and brightness of an image. For instance, the dark image has its histogram gathering in the low value of intensity ranges [17], while the brighter image has the histogram located on the high value of the intensity ranges.

Thresholds are set for histograms which have been collected into bins from 0 to 255 in each channel of HSV space, in order to select the most concentrated range of the histogram. New histograms [18] would be generated by normalizing the selected regions. Two-dimensional HS histogram features in 10×10 bins followed by ten components for V channel further consist of the final histogram feature for the classifications.

In [4], the 2D HS histograms are adopted. The color images are in HSI color space; however, the intensities are not making contributions in the HS histogram. Images taken in different parts of the GI tract appearing in slightly different color tone, and hence concentrate on different regions in the HS histogram space.

3.3 Texture Features

3.3.1 Local Binary Pattern

Besides color features, texture features are another kind of frequently employed approaches in medical image processing and local binary patterns (LBP) [19] is one of them. It was developed for texture classification by describing the differences of the local neighbors and the center pixel intensity, and designating binary values to them. The spatial structures can be measured by detecting the "uniform" patterns which are the fundamental properties of the local texture. In this approach, an image will be divided into small cells in size $N + 1$ pixels, such as 3×3 pixels. There are eight (N) neighbor pixels ($p_i, i = 0, \cdots, 8$) around the center pixel (p_c) which are compared with the center pixel. The binary values are designated to the neighbor pixels, and the center pixel is the threshold of them. That is to say, assign 1 to those neighbors equal to or greater than the center, while assign 0 to those less than the center.

$$b_i = \begin{cases} 1, & p_i \geq p_c \\ 0, & p_i < p_c \end{cases} \tag{3.7}$$

Each neighbor p_i will be transformed into an N-bit number 2^i factored by its binary value. The LBP value is then calculated for the center pixel by adding up b_i of its N neighbors.

$$\mathrm{LBP} = \sum_{i=0}^{N-1} b_i 2^i \tag{3.8}$$

From Eq. (3.8), we have that the output of LBP value depends on the starting pixel in the local neighborhood. For example, if the local neighbor traversed in the pattern of $(11100000)_2$, its LBP value is consequently calculated as$1 \times 1 + 1 \times 2 + 1 \times 4 + 0 \times 8 + 0 \times 16 + 0 \times 32 + 0 \times 64 + 0 \times 128 = 7$. However, if the binary string has been traversed as $(00000111)_2$, then the output of the LBP is then as$\mathrm{LBP} = 0 \times 1 + 0 \times 2 + 0 \times 4 + 0 \times 8 + 0 \times 16 + 1 \times 32 + 1 \times 64 + 1 \times 128 = 224$. Therefore, in the work [19], a unique pattern is assigned to identify the invariant local pattern, as below:

$$\mathrm{LBP}^{\mathrm{ri}} = \min\{\mathrm{ROR}(\mathrm{LBP}, i) | i = 0, 1, \ldots, N-1\} \tag{3.9}$$

ROR is the operator of the binary number that is circling one bit to the right at one time and N times in all. There are 36 invariant patterns out of $2^8 = 256$ in total, for a neighborhood in size 3×3.

Furthermore, the "uniform" patterns are defined and used to measure the binary pattern. First, a measurement U quantifies the times of the 0/1 transitions taking place in the binary string. For all zeros or all ones patterns, the U value is 0. For those having U value equal to 2, the "uniform" patterns are hence detected out.

$$\mathrm{LBP}^{\mathrm{riu2}} = \begin{cases} \sum_{i=0}^{N-1} b_i, & \text{if} \quad U(\mathrm{LBP}) \leq 2 \\ N+1, & \text{otherwise} \end{cases} \tag{3.10}$$

where $^{\mathrm{rui2}}$ means rotation invariant and the uniformity is no more than 2. Also, the uniformity $U(\mathrm{LBP})$as below:

$$U(\mathrm{LBP}) = |b_0 - b_{N+1}| + \sum_{i=1}^{N-2} |b_i - b_{i+1}|. \tag{3.11}$$

The histograms are accumulated on the uniform local binary patterns' outputs and form the texture features. The approach of LBP has some advantages. Because the center pixel is the binary threshold of pixels in the local neighborhood, the LBP operator outputs will not be affected by any monotonic transformation on the gray scale. So does the property against texture rotations. Last but not the least; histograms are not collected on the non-uniform patterns, given the concern that these patterns reflect the noise in the image.

LBP can also be used on RGB color image [20] by applying the operators on every color channel and their opponent channels. Therefore, the LBP texture feature of a neighborhood includes components of six histograms as a single distribution in one-dimension. In the work of [21] by Connah and Finlayson, not only histograms of LBP values in each individual channel are concatenated, but a joint histogram of three-color's distribution is calculated as well, considering the connections among channels.

LBP [19] and its extended approaches [20, 21] are adopted in analysis of WCE images as the methods of extracting texture feature, which reveals the edge information by the "uniform" measurement. In [4], both independent LBP or joint LBP histogram in RGB color space are used as texture features. And in [11], the texture features of GI tract images are extracted by the gray-scale LBP histogram operating on the I channel in HSI color space and followed by taking statistical measurements of the histogram. In [22], rotation invariant LBP is used after ulcer image has taken curvelet transforms. In those works, classifications methods such as neural networks and support vector classifiers are used after feature extraction. And hence they return the classification result of which part in GI tract the images were taken or whether bleeding/ulcer were detected in the images.

3.3.2 Gabor Filter

Textures provide a lot of discriminative information of an image. Texture features are greatly used in tasks of image classification. WCE images are relatively simple comparing with most other real-world images. Therefore, texture descriptors are representing the textures of the WCE images, such as LBP and 2D filter banks which transform images into the frequency domain.

Gabor filter is a natural way instead of setting a hard threshold on image segmentations [13]. It is a frequently used linear transformation which views images in frequency and orientation in the way of close to human visual system reaction to the image [23]. The family of 2D Gabor functions $g(x, y)$ and their frequency response of filters $G(u, v)$ are as follows:

$$\begin{aligned} g(x,y) =& \exp\left\{-\pi\left[(x-x_0)^2a^2 + (y-y_0)^2b^2\right]\right\} \\ &\times \exp\{-2\pi j[u_0(x-x_0) + v_0(y-y_0)]\} \end{aligned} \tag{3.12}$$

$$\begin{aligned} G(u,v) =& \exp\left\{-\pi\left[\frac{(u-u_0)^2}{a^2} + \frac{(v-v_0)^2}{b^2}\right]\right\} \\ &\times \exp\{-2\pi j[x_0(u-u_0) + y_0(v-v_0)]\} \end{aligned} \tag{3.13}$$

The spatial frequency is $F = \sqrt{u_0^2 + v_0^2}$ and the orientation of the filters is$\theta = \tan^{-1}\frac{v_0}{u_0}$. Literatures such as [7, 24, 25] have used Gabor filter bank to extract texture features. In [7], the final feature of an image is the superposition of all the response of the filter bank. Patterns of juices area can be distinguished by the segmentations on the response images. In [24], the complete features are concatenated by each of the filter response of the Gabor filter bank along with other features, namely, means of color channels in HSI space, and energies of high frequencies of the image in Fourier domain.

3.3.3 Scale-Invariant Feature Transform

Scale-invariant feature transform, also known as SIFT, is one of the most widely used feature for object recognition and keypoints detection/matching. The features are generated by computations of four stages [26]. Scale-space extreme detection is the first stage of identify those potential scale-invariant points which is implemented over the entire image and all scales by a difference-of-Gaussian function. In this stage, the image $I(x, y)$ is smoothed by a variable-scale Gaussian$G(x, y, \sigma)$, and the difference-of-Gaussian is produced by the subtraction of two adjacent Gaussian images as in Eq. (3.14):

$$\begin{aligned} D(x,y,\sigma) &= G(x,y,k\sigma) \times I(x,y) - G(x,y,\sigma) \times I(x,y) \\ \text{where} \quad G(x,y,\sigma) &= \frac{1}{2\pi\sigma^2}\exp\left\{\frac{(x^2+y^2)}{2\sigma^2}\right\} \end{aligned} \tag{3.14}$$

And let us denote the Gaussian smoothed image as: $L(x, y, \sigma) = I(x, y) \times G(x, y, \sigma)$.

Next, the key point candidates are further determined by performing a detailed fit which eliminates points with low contrast and poorly located on edges [26, 27].

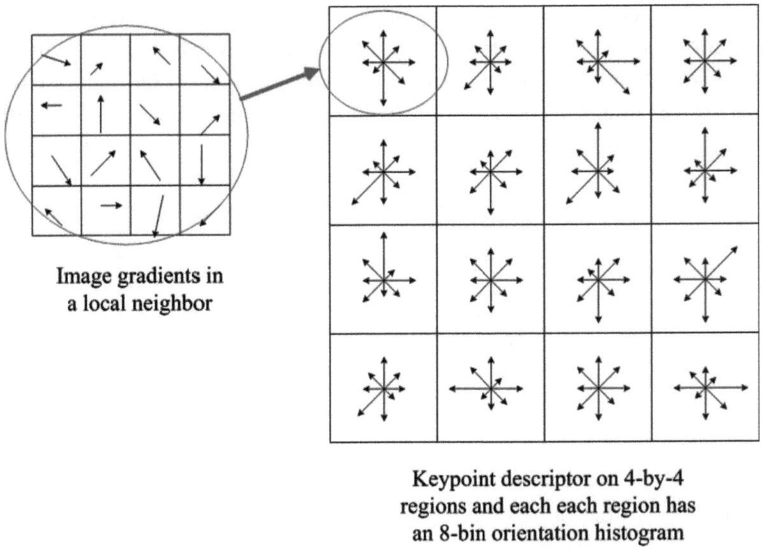

Fig. 3.3 Keypoint descriptor

The maximum and minimum pixels are found by comparing a pixel and its eight neighbors in each pyramid level and repeatedly to the level above.

The following stage is orientation assignment for the key points. The gradient magnitude $M(x, y)$ (3.15) and the gradient orientation $\theta(x, y)$ (3.16) are calculated on the Gaussian smoothed image$L(x, y)$.

$$M(x,y) = \sqrt{(L(x,y) - L(x+1,y))^2 - (L(x,y) - L(x,y+1))^2} \tag{3.15}$$

$$\theta(x,y) = \arctan \frac{L(x,y+1) - L(x,y)}{L(x+1,y) - L(x,y)} \tag{3.16}$$

The orientations of pixels in the region around a key point is then distributed into a histogram and weighted by each gradient magnitude. The dominant orientation is then determined as the peak in the histogram.

At last, the key point descriptor is finalized by the orientation histogram. In each small sample region, the orientations of each pixel are collected into an 8-bin histogram. For each key point, a typical descriptor is a 128-dimension vector of 4×4 sample regions with their 8-bin orientation histogram (Fig 3.3).

3.4 Classification Methods

3.4.1 Distances Measurements

Distance between two vectors describes the difference in a certain space. L_1 distance is used to describe the summation of the absolute differences between two vectors. It is also known as Manhattan distance or taxicab geometry. In general, the distance in ℓ_1 norm of two vectors $\mathbf{x}$ and $\mathbf{y}$ can be written as follows:

$$D_{L_1}(\mathbf{x},\mathbf{y}) = \sum_i |x_i - y_i| \tag{3.17}$$

L_1 distance usually used to discover the distance of their statistical properties between a test data and a class [16], and hence the classification decision is made based upon the distance.

The Euclidean distance is another most popular measurement, which is also called L_2 distance or ℓ_2 norm.

$$D_{L_2}(\mathbf{x},\mathbf{y}) = \sqrt{\sum_i (x_i - y_i)^2} \tag{3.18}$$

In one-dimension space, where $\mathbf{x}$ and $\mathbf{y}$ are scalar numbers, L_2 distance is just the same as the L_1 distance which is the absolute difference between two points.

Bhattacharyya distance is conducted from Bhattacharyya coefficient, which is a metric of the overlap of two statistical distributions [28]. For discrete probability distributions $\Pr(x)$ and$\Pr(y)$, their Bhattacharyya coefficient is as below:

$$\mathrm{BC} = \text{Bhattacharyya coefficient} = \sum_i \sqrt{\Pr(x_i)\Pr(y_i)} \tag{3.19}$$

For continuous probability distributions $F_x(t)$and$F_y(t)$, similarly, BC can be defined as below:

$$BC = \int \sqrt{F_x(t)F_y(t)}\mathrm{d}t \tag{3.20}$$

The Bhattacharyya distance hence has the definition as:

$$D_{\mathrm{Bhattacharyya}} = -\ln(\mathrm{BC}) \tag{3.21}$$

Use distances difference to classify descriptors, is a relatively intuitive and simple calculation method, such as what used in the past works [16, 24]. In [16], in order to give a justified evaluation without strong classifier's effects on the descriptors themselves, ℓ_1 and ℓ_2 norm distance are used to quantify them. And in [29], the distance of local color histogram is calculated by L_1 distance between two image patches $\mathbf{x}$ and $\mathbf{y}$ as below, where $H_{r,i}^x$ is the histogram of patch x for red color channel (r) in the ith bin, and so on so forth.

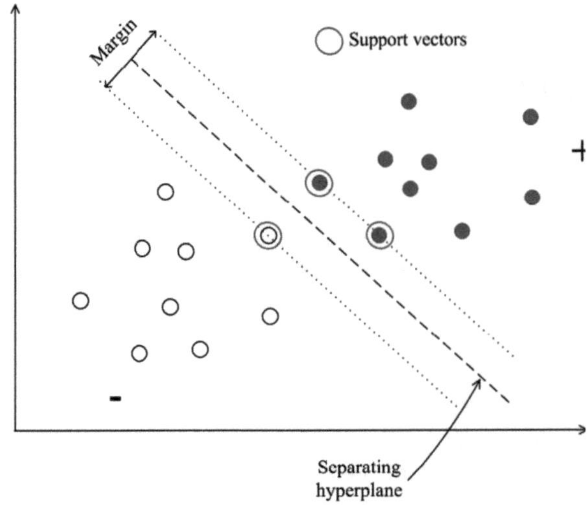

Fig. 3.4 The support vectors are composing a pair of hyperplanes, and the shortest distance between them is defined as the "margin". In linear separable data with binary classes, the separating hyperplane is then determined by maximizing the margin

$$D_{L1}(\mathbf{x},\mathbf{y}) = \sum_{i=1}^{N_{\text{Bins}}} \left(|H^x_{r,i} - H^y_{r,i}| + |H^x_{g,i} - H^y_{g,i}| + |H^x_{b,i} - H^y_{b,i}| \right) \tag{3.22}$$

To evaluate the similarity of the two histogram vector features, in [5, 24], measurement is modified based on the Bhattacharyya distance. First, the distance $D_{\text{Bhattacharyya}}(f_i, f_j)$ is conducted between any two frame features f_i andf_j. Instead of measuring any two consecutive frames, where the sudden change might be brought by transmission noises, the measurement of the ith frame is an average of a collection of the distances in its following sequence.

$$C(i) = \frac{1}{9} \sum_{k=i}^{i+2} \sum_{j=3}^{5} D_{\text{Bhattacharyya}}(f_k, f_{i+j}) \tag{3.23}$$

3.4.2 Support Vector Machine

Support vector machine (SVM) belongs to the discriminative model categories in the machine learning. It was first introduced by Cortes and Vapnik in 1995 [30, 31]. The initial intension of the SVM was to find out the best separation between two classes of linear separable data, where the best separation can be defined as equal distance from both sides of the data. A pair of parallel hyperplanes consists of support vectors which refer to those data samples that are closest to the other class. And the distance between these two parallel supporting hyperplanes are called "Margin", as shown in Fig. 3.4.

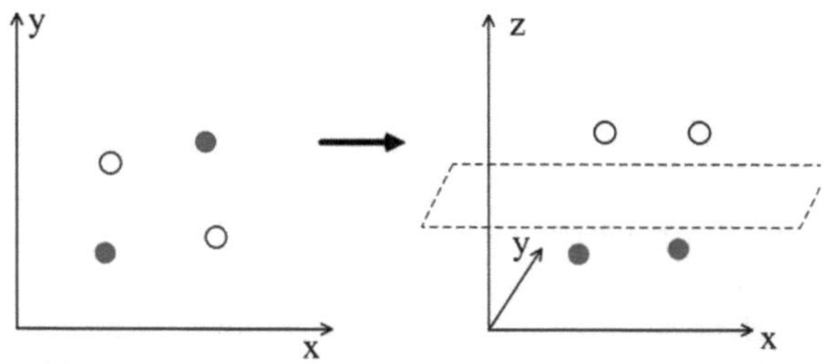

Fig. 3.5 Non-separable data could be mapped into a higher dimensional space to be separated

Table 3.1 Examples of commonly used kernel function in SVM

Polynomial function	$K(x_i, x_j) = (x_i \cdot x_j + 1)^p$
Gaussian radial basis function (RBF)	$K(x_i, x_j) = \exp(-\|\|x_i - x_j\|\|^2/2\sigma^2)$
Two-layer sigmoidal neural network	$K(x_i, x_j) = \tanh(\kappa x_i \cdot x_j - \delta)$

Typically, SVM is designed for binary classification problems. For multiple classes' data, usually the classifiers are trained to distinguish one class against rest of the other classes, also known as "one-against-all" [32]. The label of a testing sample is decided by the highest output of the decision function. Another mostly used approach, "one-against-one" strategy [32], requires $\frac{M(M-1)}{2}$ classifiers to classify M labels of data, and each classifier is trained for any two classes among all. And then the testing data are determined by the most voted label after all. The directed acyclic graph SVM (DAGSVM) is also used based on "one-against-one" training strategy, but the label is moved along a path on the graph based on the output of each binary SVM classifier [32, 33].

A separate hyperplane may not be found, because the data are not linearly separable. To solve this problem, kernel functions which map data into a higher dimensional space are introduced into SVM, so that the projected data could be separable, such as in Fig. 3.5. There are some examples of kernel function$K(x_i, x_j)$, listed here in Table 3.1.

SVM is a classification machine that guarantees the optimal solution to obtain the minimized error from the theoretical perspective. It is one of the most widely applied to various classification problems. In [14], Charisis et.al utilized SVM for the ulcer detection in the WCE images. First, bi-dimensional Empirical Mode Decomposition was applied to remove noises in the raw WCE images, and then the lacunary features were exacted followed by the SVM classification. The experimental results demonstrated that SVM can separate the ulcer images with higher classification accuracy.

3.4.3 Artificial Neural Networks

Artificial Neural Networks (ANN) is the discriminative method for machine learning. Inspired by the biological learning system that is connected by the complex neurons, Alexander Bain (1873) and William James (1890) [34] proposed the preliminary neural network model. Since then, different learning methods, such as Hebb learning and back propagations have been introduced to neural networks and proven to be successful in many real-world applications. ANN consists of different layers, such as input layer, hidden layer, and output layer. Each layer includes a number of nodes (neurons), all the nodes are either fully connected or partially connected inside each layer as well as between different layers. Each node is associated with a fire function, which controls the state of this node, i.e., the output to the connected nodes. In general, artificial neural network can be applied to approximately any real, discrete, and vector valued target functions, which makes it very popular and successful in various applications. However, unlike SVM, ANN lacks the theoretical analysis of learning process. Besides the heavy computational cost, the training of ANN may be prone to over fitting due to its "black box" nature. Except adjusting the iteration number, learning rate and kernel functions, it may be difficult to improve the performance of neural networks.

In [14], Kodogiannis and Boulougoura proposed a neural-network-based classification approach for WCE image classification. Texture feature was extracted from both normal and abnormal images, and then extended normalized radial basis function network was utilized for classification. The experimental result shows that the proposed approach can achieve competitive classification accuracy with the texture unit number features.

3.4.4 Hidden Markov Models

Hidden Markov Model (HMM) is a statistical Markov model, in which the system state is a Markov process and unobservable [43]. HMM has been successfully applied in the speech recognition, gesture recognition, and many other temporal sequential recognition applications. HMM can be treated as the dynamic Bayesian network in a simple form. The transition matrix and confusion matrix represent the state transition probability and the observation probability conditional on each hidden state. Given the training data set, the training process will learn and determine the value of these two matrixes. After training, for a test sequence of observations, the trained HMM will give the corresponding hidden state sequence with the highest probability. Many algorithms, such as forward, backward propagation, and Vitebi have been implemented in the HMM model. In [4], the WCE video was divided to four hidden states according the gastrointestinal tract. HMM is then trained and evaluated based on the observation, i.e., the video frames.

3.5 Medical Applications

3.5.1 Video Segmentation

Wireless capsule image has been widely used for diagnosis in medicine since 2001. After the patient swallowed the pill-shaped capsule, it will take up to 8 h for a patient to record the images inside the gastrointestinal tract (GI). As the total number of images stored in the capsule is around 50,000 (2–3 h videos), it is a tedious and time-consuming task for a doctor to go through all the images, which leads to many research work [4, 5, 10, 35] contributing to the wireless capsule video analytics.

Intuitively, the first attempt of speeding up long video analytic is to split it into several video clips, similar to the event detection in the semantic video analysis. Based on the gastrointestinal tract structure, previous work [35] suggested that the WCE video can be divided into four meaningful parts: entrance, stomach, intestine, and colon, which are corresponding to different routes in the GI tract. In particular, in order to facilitate the localization function in the RAPID software [36], it is very crucial to mark two important locations in the WCE videos: Pylorus (the valve between stomach and the intestine) and Ileocecal valve (between the intestine and colon). Accurately localizing these two valves appears to be a difficult task even for experienced clinicians, as the images in these boundary areas look very similar to each other. Different video segmentation approaches have been introduced in recent years, and most of them can be categorized into two big groups as detailed in the following: classification-based approach and key frame detection-based approach.

3.5.1.1 Classification-Based Approach

From the classification perspective, all the images in the WCE video can be assigned with a class label in terms of four different regions in the GI tract: entrance, stomach, intestine, and colon. After manually annotating the class label for the training image, the WCE video segmentation can be treated as a typical classification problem. A classifier will be built upon these labeled images and can be used to identify any new image frames. Among these classifiers, ANN and SVM are the most popularly used tools for WCE video segmentation.

In [35], the MPEG-7-based descriptor is used to perform the classification of these four zones. Bayesian classifier with different prior distribution was evaluated and compared, and the experiment results indicated that the classifier based upon L_1 norm and the classifier with nonlinear prior probability could archive the best classification performance with regard to the scalar color features.

In [4], Mackiewicz et.al investigated different classification methods based upon a number of features, such as compressed histogram, color and texture, subimage region, and motion. Their experiments indicated sliding window and

HMM classifiers based upon combined global and local features outperformed other classifiers.

3.5.1.2 Key Frame Detection-Based Approach

Key frame detection has been widely applied to different video segmentation tasks for video content analysis. Usually, a sliding window along the temporal direction is used to scan the whole video step-by-step, and a target function is evaluated on the extracted feature from all the consecutive video clips. The abrupt change or key frame is determined when the target value is above a certain threshold. In the WCE video segmentation, the key frame detection is mainly to find the boundaries of the digestive organs, i.e., the pylorus (the valve between stomach and intestine) and the ileocecal (the valve between intestine and colon).

In [5, 37], Gallo et.al first divided the entire videos to short video frame sequences with each one containing six frames. The texton feature was extracted from each frame and used to build a texton dictionary. For each short video sequence, a texton histogram was formed based upon the trained dictionary. To search the key frame or abrupt change, the similarity between two neighboring short sequences was evaluated by Bhattacharyya distance measure. Their experiment indicated their method can greatly reduce the time expense with a low miss rate. Vu et.al [10] also developed dominant color with KNN clustering method to detect the boundary of two adjacent parts in the WCE videos.

3.5.2 Feature-Based Abnormal Detection

Early research on automatic detection of abnormal cases is proposed by Kodogiammis in [14] and in [38] with Boulougoura. In the feature extraction stage, nine features are taken from each histogram of six color channels in RGB and HSV color space. These nine features are the statistical measurements including standard derivation, variance, skew, kurtosis, entropy, energy, inverse difference moment, contrast, and covariance. In addition, a scheme of texture unit number was considered in [14], which obtained the histogram based on texture units in the manner that is very similar to local binary pattern (LBP) scheme. And those features are hence estimated based on the histograms of the texture units over the entire image. In the classification stage, artificial neural networks are used for training and classifying features from each color plane. The schemes of fuzzy inference neural network (FINN) [38] and extended normalized radial basis function network (ENRBF) [14] are proposed. The diagnose decisions are made on incorporating decisions of all six subsystem (R, G, B, H, S, and V), namely, the fuzzy integral scheme, instead of classical winner-takes-all decision strategy. In the experiments of [38], there are 73 images in the image dataset, including 25 normal and 23 abnormal images as training samples, and after fused the output decisions, the classification accuracy is 100 % with confident level ranging from 0.64 to 0.98. In [14], there are 35 training images of normal and abnormal case, respectively, and same as the testing set. The overall accuracy is 95.71 %. The performance from the experiments is relatively good; however, their image database is very limited.

A recent work of bleeding detection was proposed in [18], where a three-level classification method has developed. Adaptive color histograms are extracted in each small cell in HSV color space and hence classified by Neural Networks (NN) as the low-level classification. In the intermediate level, NN classifier is again used on blocks which consist of 3-by-3 cells. At last, the final decision of a block is made upon previous two levels of classification results based on a set of thresholding rules. This approach is different from most other papers in the way of making classifications block by block and further regionalizing the bleeding area in a WCE image. The testing results perform better compared with other histogram features as well as results just based on low-level classification. However, the size of the testing dataset is not specified in this paper.

3.5.3 Region-Based Abnormal Detection

Intuitively, endoscopy images have rather simpler scenarios than other natural images taken in live. A common type of approach is to segment images into non-rigid regions based on their color or texture properties and distinguish the local areas between abnormal symptoms and normal intestinal walls by certain thresholds.

In segmentation stage, color spaces themselves could simply provide much information. In [12], images have transformed into HSI color space, and information from each channel is connected under the proposed segmentation scheme. After segmentations, an extended minimum transform which is as a thresholding operation is presented to obtain the outline of the regions, and the number of the regions of an image decide whether it is an abnormal case. This proposed approach has tested on 50 images in normal and abnormal cases, respectively, and finally 92 and 96 % accuracies have achieved.

Intensity-based features and color features are discussed by Lau and Correia in [39, 40]. Local contrast, Sobel edge operator-based edge matching feature, a dual threshold method, and entropy-based similarity feature have been utilized on grayscale images. On color images, features are extracted from RGB and HSV spaces. Namely, mutual information and thresholds are considered on R, G, and B channels simultaneously, and red color is thresholded in between values of 140 and 255 obviously detecting the blood. Last but not least, HSV color space allows recognizing color pattern of bleeding area in high saturation and Crohn's diseases which is with low saturation as well as increased hue values. With implemented in those eight features, thresholds of pixel intensities are studies on abnormal cases. In an image with bleeding areas, the dual thresholds of pixel intensities are 80 and

Table 3.2 Number of frames in each video

Video	Number of Frames
Bleeding	509
Crohn's disease	440
Polyp	302
Normal	851

100, which, for the Crohn's disease, are 120 and 200. Further classification experiments on bleeding cases then have done in [40], and correct detection rate reaches 88.3 % as well as with false and missed detections.

In [41], Karargyris and Bourbakis have proposed a method on blood abnormality detection. WCE images are first taken the Karhunen–Loeve (K–L) transformations. Each RGB channel is weighed by the eigenvectors obtained from the covariance matrix of the channels. Next, segmentations are implemented by the method of growing seeds based on transformed K–L color space. Finally, local regions are integrated based on global topological relationships. Software has developed based on their proposed methodology to register the bleeding regions.

Besides information directly from color space, texture information obtained by Gabor filters is also applies to segment images into binary regions in order to classify juices and bubbles occlusion [7]. While the intestinal is covered with juices and bubbles, it does not bring much information to diagnose. But detections on them could reduce the non-valid frames and hence save the overall time consumption. Experimented on ten videos with around 20,000 frames, there is roughly 20 % of time reduced.

A recent work [25] has also adopted Gabor filter to do the segmentations, followed by K-means clustering method to obtain the levels of elevation. The polyp regions are selected by curvature centers ratio. A relatively small testing dataset with 64 polyp images and 64 normal images is used on testing their proposed method. And when the clustering factor K is between 7 and 11, the correct identification rate on polyp images is 100 %, as well as 19 % of the false positive rate.

3.6 Experiments

Experiments are used to verify the efficiency and accuracy of the abnormal case identification. Video data are obtained from the website of SynMed [42]. There are three lesion classes in the tests, including bleeding, Crohn's disease, and polyps, besides of which, a normal case is the forth class. Video frames are 320 $\times$ 320 pixels and each case is in a video clip with number of frames from 300 to 800, as shown in Table 3.2.

There are two kinds of most popular features, dense-SIFT and LBP, extracted from each image to characterize the local regions. In our experiments, dense-SIFT

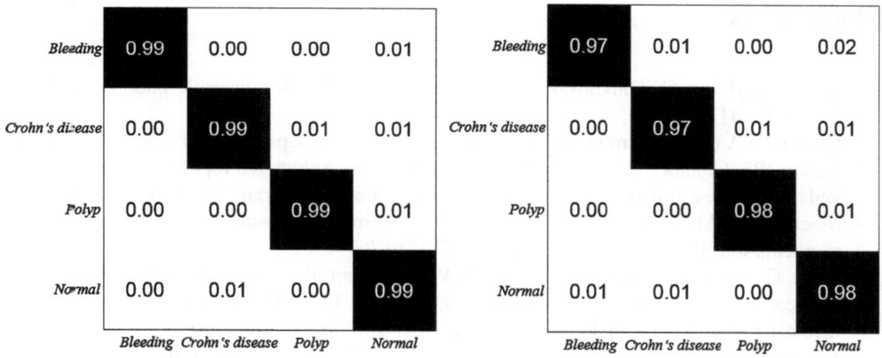

Fig. 3.6 Confusion matrix of the classifications with features of dense-SIFT (*left*) and LBP (*right*)

is effected on each block in size of 160 × 160 pixel and blocks are overlapped with half of the block size. Each block yields a dense-SIFT histogram as 128 dimensions of vector. In order to describe the entire image thoroughly while preserving each regional gradient, vectors of blocks are concatenated into one to be classified. On the contrary, LBP is a 59-bin histogram which contains the information of the entire image.

In the experiments, we used linear-SVM as the classifier, and divided data into 10 groups evenly in each class so that the results could be cross-validated. Among those data, there are only 10 % of samples in each class that are used for training the classifier, while the rest is used as the testing data. Although, there are small amount of training samples, the accuracies are very reliable, achieving 97 % in general. Figure 3.6 shows the confusion matrix of the classification result on each class.

3.7 Future Direction

The WCE technology has now become an irreplaceable method on navigating the human GI tract and assisting doctors in diagnosing the symptoms. More and more companies around the world have begun to study and invent even more advanced techniques of WCE. Doctors would require more real-time interactions with the WCE by remotely controlling it, such as staying in a certain region, zooming into focus on a particular area, moving backwards and releasing medications on exact lesion point, without bringing any difficult time to the patients.

Acknowledgments Partial support for this work was provided by the National Science Foundation's Course, Curriculum, and Laboratory Improvement (CCLI) program under Award No. 0837584. Any opinions, findings, and conclusions or recommendations expressed in this material are those of the authors and do not necessarily reflect the views of the National Science Foundation.

References

1. Iddan G, Meron G, Glukhovsky A, Swain P (2000) Wireless capsule endoscopy. Nature 405(6785):417
2. Qureshi WA (2004) Current and future applications of the capsule camera. Nature 3:447–450
3. Cunha JPS, Coimbra M, Campos P, Soares JM (2008) Automated topographic segmentation and transit time estimation in endoscopic capsule exams. IEEE Trans Med Imaging 27(1):19–27
4. Mackiewicz M, Berens J, Fisher M (2008) Wireless capsule endoscopy color video segmentation. IEEE Trans Med Imaging 27(12):1769–1781
5. Gallo G, Granata E, Scarpulla G (2009) Wireless capsule endoscopy video segmentation, medical measurements and applications (MeMeA). In: IEEE international workshop on medical measurements and applications, pp 236–240, 29–30 May 2009
6. Karargyris A, Bourbakis N (2011) Three-dimensional reconstruction of the digestive wall in capsule endoscopy videos using elastic video interpolation. IEEE Trans Med Imaging 30(4):957–971
7. Vilarino F, Spyridonos P, Pujol O, Vitria J, Radeva P, de Iorio F (2006) Automatic detection of intestinal juices in wireless capsule video endoscopy. In: 18th international conference on pattern recognition, vol 4, pp 719–722
8. U.S. FDA (2012) Given\textregistered diagnostic imaging system—K010312. http://www.fda.gov/MedicalDevices/ProductsandMedicalProcedures/DeviceApprovalsandClearances/Recently-ApprovedDevices/ucm085396.htm. Accessed 6 Feb 2012
9. Charisis V, Hadjileontiadis, LJ, Liatsos CN, Mavrogiannis CC, Sergiadis GD (2010) Abnormal pattern detection in Wireless capsule endoscopy images using nonlinear analysis in RGB color space. In: IEEE engineering in medicine and biology society (EMBC), pp 3674–3677, 31 Aug 2010–4 Sept
10. Vu H, Yagi Y, Echigo T, Shiba M, Higuchi K, Arakawa T, Yagi K (2010) Color analysis for segmenting digestive organs in VCE In: 20th international conference on pattern recognition, pp 2468–2471, 23–26 Aug 2010
11. Li B, Meng MQ-H (2009) Computer-aided detection of bleeding regions for capsule endoscopy images. IEEE Trans Biomed Eng 56(4):1032–1039
12. Dhandra BV, Hegadi R, Hangarge M, Malemath VS (2006) Analysis of abnormality in endoscopic images using combined hsi color space and watershed segmentation. In: 18th international conference on pattern recognition, ICPR 2006, vol 4, pp 695–698
13. Karargyris A, Bourbakis N (2011) Detection of small bowel polyps and ulcers in wireless capsule endoscopy videos. IEEE Trans Biomed Eng 58(10):2777–2786
14. Kodogiannis VS, Boulougoura M (2005) Neural network-based approach for the classification of wireless-capsule endoscopic images. In: IEEE international joint conference on neural networks, vol 4, pp 2423–2428, 31 July 2005–4 Aug 2005
15. Mackiewicz M, Berens J, Fisher M, Bell D (2006) Color and texture based gastrointestinal tissue discrimination. In: IEEE international conference on acoustics, speech and signal processing, pp II–II, Toulouse, May 2006
16. Coimbra MT, Cunha JPS (2006) MPEG-7 visual descriptors contributions for automated feature extraction in capsule endoscopy. IEEE Trans Circuits Syst Video Technol 16(5):628–637
17. Gonzalez RC, Woods RE (2008) Digital image processing, 3rd edn. Pearson/Prentice Hall, Upper Saddle River
18. Chee K, That M, Li L, Shen W, Liu J, Joo H, Chan K, Tan P (2010) Multi-level local feature classification for bleeding detection in wireless capsule endoscopy images. In: IEEE conference on cybernetics and intelligent systems (CIS), pp 76–81, 28–30 June 2010
19. Ojala T, Pietikainen M, Maenpaa T (2002) Multiresolution gray-scale and rotation invariant texture classification with local binary patterns. IEEE Trans Pattern Anal Mach Intell 24(7):971–987

20. Maenpaa T, Pietikainen M (2004) Classification with color and texture: jointly or separately? Pattern Recogn 37:162940
21. Connah D, Finlayson, GD (2006) Using local binary pattern operators for color constant image indexing. In: Third European conference on color in graphics, imaging and vision
22. Li B, Meng MQ-H (2009) Texture analysis for ulcer detection in capsule endoscopy images. Image Vis Comput 27(9):1336–1342
23. Daugman JG (1988) Complete discrete 2-D Gabor transforms by neural networks for image analysis and compression. IEEE Trans Acoust Speech Signal Process 36(7):1169–1179
24. Gallo G, Granata E, Scarpulla G (2009) Sudden changes detection in WCE video. In: International conference on image analysis and processing, Vietri sul Mare
25. Hwang S, Celebi ME (2010) Polyp detection in wireless capsule endoscopy videos based on image segmentation and geometric feature. In: 2010 IEEE international conference on acoustics speech and signal processing (ICASSP), pp 678–681, 14–19 March 2010
26. Lowe DG (2004) Distinctive image features from scale-invariant keypoints. Int J Comput Vis 60(2):91–110
27. Lowe DG (1999) Object recognition from local scale-invariant features. In: The proceedings of the seventh IEEE international conference on computer vision, vol 2, pp 1150–1157
28. Kailath T (1967) The divergence and Bhattacharyya distance measures in signal selection. IEEE Trans Commun Technol 15(1):52–60
29. Vu H, Echigo T, Sagawa R, Yagi K, Shiba M, Higuchi K, Arakawa T, Yagi Y (2006) Adaptive control of video display for diagnostic assistance by analysis of capsule endoscopic images. In: International conference on pattern recognition, vol 3, pp 980–983
30. Cortes C, Vapnik V (1995) Support-vector networks. Mach Learn 20(3):273–297
31. Burges CJC (1998) A tutorial on support vector machines for pattern recognition, data mining and knowledge discovery, vol 2, no 2, pp 121–167
32. Hsu C, Lin C (2002) A comparison of methods for multiclass support vector machines. IEEE Trans Neural Netw 13(2):415–425
33. Duan K, Keerthi SS (2005) Which is the best multiclass SVM method? An empirical study. In: Proceedings of the sixth international workshop on multiple classifier systems, pp 278–285
34. Bishop CM (1996) Neural networks for pattern recognition. Oxford University Press, Oxford
35. Coimbra M, Campos P, Cunha JPS (2005) Extracting clinical information from endoscopic capsule exams using MPEG-7 visual descriptors. In: The 2nd European workshop on the integration of knowledge, semantics and digital media technology, pp 105–110, London, Nov 2005
36. RAPID software (2012) Capsule endoscopy software by given imaging. http://www.givenimaging.com/en-us/HealthCareProfessionals/Products/Pages/Software.aspx. Accessed 6 Feb 2012
37. Gallo G, Granata E (2010) WCE video segmentation using textons. In: Proceedings of SPIE medical imaging, San Diego, Feb 2010
38. Kodogiannis VS (2004) Computer-aided diagnosis in clinical endoscopy using neuro-fuzzy systems. In: IEEE international conference on Fuzzy systems, vol 3, pp 1425–1429, 25–29 July 2004
39. Lau, PY, Correia PL (2007) Analyzing gastrointestinal tissue images using multiple features. In: Proceedings of conference on telecommunications, Peniche, vol 1, pp 435–438, May 2007
40. Lau PY, Correia PL (2007) Detection of bleeding patterns in WCE video using multiple features. In: 29th annual international conference of the engineering in medicine and biology society, pp 5601–5604, 22–26 Aug 2007
41. Karargyris A, Bourbakis N (2008) A methodology for detecting blood-based abnormalities in wireless capsule endoscopy videos. In: 8th IEEE international conference on bioinformatics and bioengineering, pp 1–6, 8–10 Oct 2008
42. Synmed (2012) Demonstration videos by capsule endoscopy. http://www.synmed.co.uk/products_capsule_endoscopy.htm. Accessed 6 Feb 2012
43. Rabiner LR (1989) A tutorial on hidden Markov models and selected applications in speech recognition. In: Proceedings of the IEEE, vol 2, no 3, pp 257–286, Feb 1989

Chapter 4
Nanotechnology Innovations for Low-Temperature Fuel Cells for Micro Autonomous Systems

Ayokunle Omosebi and Ronald S. Besser

Abstract The requirement for longer and uninterrupted operation by micro autonomous systems calls for alternatives to the incumbent battery technology. The proton exchange membrane fuel cell (PEMFC) presents a comparative advantage over battery technology due to their higher energy density and ability to operate continuously without the need for recharging. The cost of expensive catalyst materials utilized by the PEMFCs has so far impeded this technology. The dramatic success of the electronics industry in making cheaper and more efficient products has created new pathways for PEMFC advancement. This chapter discusses the integration of nano/microfabrication practices and techniques to fuel cell systems design. A new technique with the ability to produce high-aspect ratio features of sub-micrometer critical dimension and larger by leveraging the tools of electron beam lithography and advanced dry etching from the established techniques of nano/microtechnology is also discussed. This capability opens the possibility of creating a variety of novel architectures for fuel cells and other electrochemical devices (including sensors, electrolyzers, and electrochemical reactors, among others) and the promise of cost reduction through the model of microelectronics technology that leverages integration of components, rapid batch processing, automation, and economies of scale.

A. Omosebi · R. S. Besser (✉)
Department of Chemical Engineering and Materials Science,
Stevens Institute of Technology, Hoboken, NJ, USA
e-mail: rbesser@stevens.edu

A. Omosebi
e-mail: aomosebi@stevens.edu

Y. Guo (ed.), *Selected Topics in Micro/Nano-Robotics for Biomedical Applications*,
DOI: 10.1007/978-1-4419-8411-1_4, © Springer Science+Business Media New York 2013

4.1 Introduction

Autonomous systems require limited or no human input to perform their designated functions. Wireless autonomous systems primarily rely on batteries as power sources which limits the functioning time of the device and leads to interrupted operation as the batteries must either be recharged or replaced [16, 31]. For uninterrupted and continuous operation, new power sources including vibration energy harvesters [5], and fuel cells [16, 31] are being considered as battery replacements. Proton exchange membrane fuel cells (PEMFCs), also regarded as polymer electrolyte membrane fuel cells, are high energy density electrochemical devices that utilize a polymer membrane usually Nafion, to facilitate ion transport between the anode and cathode parts of the cell. PEMFCs are attractive for their low emissions, quiet operation, scalability, and high efficiency [2, 28]. Fuel cells are fundamentally similar to batteries as they are both electrochemical devices that produce DC electricity. However, with fuel cells, the sources of both the fuel and oxidant are external to the system thus allowing for longer operation intervals. The wide-scale adoption of PEMFCs as an alternative is however currently limited by the cost of materials used in their construction [12].

The rapid growth of the electronics industry over the past five decades has been heralded by advances in microfabrication technology, miniaturization, and rapid prototyping. As evidence, today we have cheaper, faster, and more efficient electronic devices. Borrowing from the success of the electronics industry, researchers have explored synergistic combinations of nano/micro fabrication and electrochemical systems toward the deployment of cost effective fuel cell systems. This has led to the investigation of micro-electro-mechanical systems (MEMSs) technology for fuel cell construction [27, 33]. The relevant MEMS technology includes techniques such as etching, micro-molding, hot embossing, sputtering, chemical vapor deposition, physical vapor deposition, nanoimprint lithography (NIL), beam-based (optical, electrons, ions) lithography, etc.

1. *Nafion Polymer Electrolyte*

Nafion® is a perfluorosulfonate ionomer made up of a hydrophobic polytetrafluoroethylene (PTFE) backbone and side chains terminated by a hydrophilic sulfonic acid group. It is permeable to water, inert to chemical attack, conductive toward cations, and selective to gas permeation [3, 9, 13, 22]. Nafion melts at approximately 135 °C, thus creating an upper limit on its application. In a PEMFC, proton conduction through Nafion is a function of water content, preferring humid over dry conditions. Under humidified conditions, Nafion is an excellent proton conductor at temperatures below 90 °C. Nafion is used for hydrogen production via the electrolysis of water, electrochemical sensing and actuation, electrolysis of sodium chloride solutions in chlor-alkali cells, drying and humidification operations, and power production in a fuel cell [7, 15, 20]. Nafion is generally attractive due to its stability in acidic and basic media, fast cation

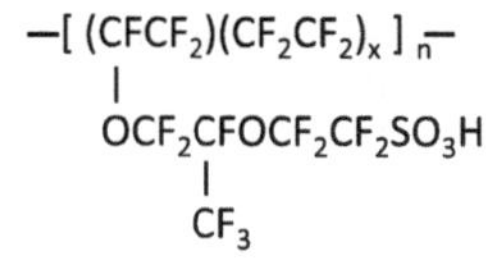

Fig. 4.1 Chemical structure of Nafion

transport, and its response to electrochemical potential. Figure 4.1 shows the chemical structure of Nafion.

In most applications involving Nafion, it is typically adjacent to a metal electrode serving as electrical connection or catalyst material. While widely known for its use in electrochemical applications, Nafion has also been explored for ionic polymer metal composite (IPMCs) applications. IPMCs are formed by coating metal electrodes usually platinum or gold onto either side of the Nafion polymer. The application of an electric field across the IPMC causes the transport of cations and water molecules to the cathode, which results in volume change and causes the membrane to bend toward the anode. These devices can be used as micropumps, actuators in artificial muscles, and microsensors. An interesting application of the Nafion IPMC is the biometric jellyfish robot which mimics the real locomotive behavior of a jellyfish [17, 29]. Likewise, the membrane electrode assembly (MEA) of a PEMFC consists of 2 electrode–Nafion interfaces which form the anode and cathode catalyst layers typical of electrochemical systems. Platinum nanoparticles are typically dispersed on carbon to maximize the availability of reaction sites for electrochemical power generation. The carbon-supported catalyst is then sprayed in the form of an ink onto Nafion to form catalyst layers approximately 10–50 μm thick.

4.1.1 PEMFC Construction and Operation

For electrochemical power generation in the PEMFC, the fuel and oxidant are fed to the anode and cathode respectively. The most commonly utilized fuels are hydrogen and methanol, while the oxidants are oxygen and air. The H_2 based PEMFCs are popularly regarded as simply PEMFCs, while the methanol-based are called the direct methanol fuel cells (DMFCs). Protons and electrons are generated at the anode of the PEMFC by the oxidation of the fuel. The generated protons are conducted via the membrane to the cathode, while the electrons are conducted through an external circuit doing work as they move to the cathode. A recombination reaction occurs at the cathode completing the process. To assemble the PEMFC, the MEA is sandwiched between two gas diffusion layers (GDLs) that serve to homogeneously redistribute the incoming gas phase reactants to the reaction plane at the Nafion-catalyst interface and to promote electronic conduction. This structure is in turn sandwiched between two graphite flow fields that supply the incoming fuel and oxidant to the appropriate GDLs. A labeled diagram of the PEMFC is shown in Fig. 4.2. A major requirement for materials used in

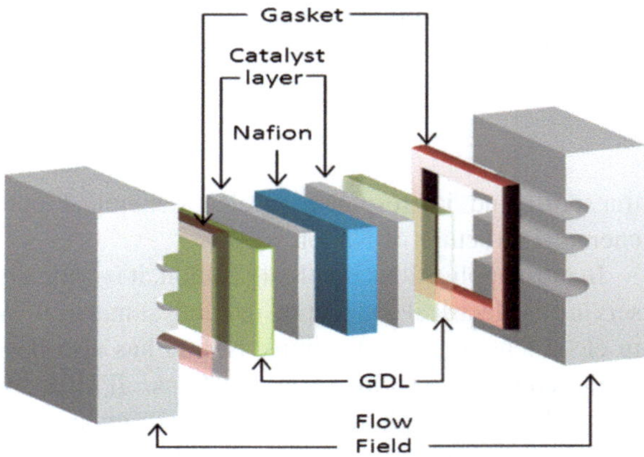

Fig. 4.2 Components of a PEMFC

constructing the PEMFC is electrochemical stability. When methanol is used as fuel, ruthenium is typically added to the catalyst to mitigate catalyst deactivation.

Despite the proliferation of material variations for PEMFCs, the gold standard of performance remains the Nafion membrane with precious metal catalyst layers. Membrane improvements and catalyst optimization have led to better performance through thinner, higher conductivity membranes and more active dispersed nanoparticle catalysts. However, the basic architecture, defined by the stacking of flat, two-dimensional membranes with current flow normal to the plane, has remained nearly unchanged since the Gemini space program of the 1960s. In large measure, the lack of disruptive resources for PEMFC technology development has resulted in only moderate performance improvements and cost reductions over a relatively long timeframe, with the result that PEMFC implementation is sadly lacking. The ability to directly pattern a Nafion membrane with arbitrary surface structures opens the possibility of new architectures which may possess the performance and cost attributes needed for rapid adoption.

4.1.2 Nano/Microfabrication Techniques for PEM Applications

In order to address the high cost of platinum catalysts, various attempts at applying nano/microfabrication have been made. Sputter deposition of catalyst layers has been investigated toward the deployment of ultra-low loading MEAs [14, 21]. Pt loadings as low as 0.054 mg/cm^2 (25 nm) have been demonstrated to yield power densities of the order of 210 mW/cm^2 (38889 mW/mg) [14] for H_2 PEMFCs. The addition of a sputtered layer of Cr, Pd, or Ti under-layer prior to Pt deposition have led to even higher performances [14, 23].

Additionally, sputter deposition from a variable deposition angle with constant rotation has been used to create ordered particle-like electrodes for PEM fuel cell applications [11]. Despite continued improvement, the performance of the sputtered catalyst MEAs is still lacking in terms of area specific power density (mW/cm^2) compared to the conventional carbon-supported ink-based MEAs. It has been reported [21] that increased thickness of sputtered Pt layers leads to transport resistance for O_2 which results in loss of cell performance.

Research has also been focused on methods for directly modifying the surface morphology of Nafion. The cathode reduction reaction requires the presence of protons, electrons, and oxidant at the reaction plane to generate power. The ease of having this combination reaction occur in turn affects the performance of the fuel cell. It is therefore necessary to increase the availability of this interface, while minimizing the transport resistance to protons, electrons, and oxidant accessing the reaction plane [21]. To achieve this objective, the membrane–electrode interface could be restructured to possess a larger interfacial area. In work done by the group of Prof. F. Prinz [18] it was found that abrasive roughening gave an increase in current density over non-roughened membrane devices. It was surmised that the modest increase in interfacial area, although highly irregular and subject to mechanical damage, resulted in performance gains. In another study [26], Nafion was treated by bombarding it with Ar^+ prior to spraying catalyst ink onto it. Pt loading was varied from 0.1 to 0.55 mg/cm^2, and Ar^+ treated versus untreated PEMFCs were compared. It was found that increasing the ion dose increased the surface roughness for which there was an optimum in terms of effect on PEMFC performance. The treated membrane exhibited better performance, higher catalyst activity, and a lower charge-transfer resistance. Other physical etching strategies have also been investigated for the surface modification toward achieving better PEMFC performance [6, 8, 22].

Investigations by various groups involving modest increases in interfacial area have also shown benefits to PEMFC performance. Zhou and co-workers [34] began with a liquid ionomer and used soft lithography molding techniques to create microscale ($\sim$1–5 μm feature size) depressions in the membrane. Interfacial surface area was increased up to just over 2× as limited by the microfabrication technique employed. Power density increased with area enhancement, by an impressive 61 % over otherwise equivalent flat membranes. However, mass transport limitations lessened performance gains at high currents, due to flooding under operating conditions at 50 °C. The use of NIL for Nafion structuring has also been reported [1, 30, 32]. NIL is an emerging integrated circuit manufacturing technique that creates micro and nanoscale features normally handled by lithography through the formation of imprinted surface structures. Zhang and co-workers [32] created shallow dot patterns 500 nm in diameter, 140 nm in depth, with 2 μm spacing on the surface of a Nafion membrane. The aspect ratio of the structure was 0.3, which provided a surface area enhancement approximately 2 % times larger than the planar surface. A 20 nm Pt layer was sputtered directly onto both sides of the membrane to create a nominal loading of 0.04 mg/cm^2. This structure was compared to a non-patterned Nafion-sputtered layer structure, to which it showed modest performance increase.

Miniature fuel cells with critical dimensions of 1 cm^2 and less have been fabricated in whole or part via MEMS's approach using silicon, metal, and polymer substrates [24]. An example of the MEMS integration process is a polydimethylsiloxane (PDMS) micro fuel cell which was fabricated by employing micromolding [27]. In order to form the nano/micropatterns in Nafion, which access a wider range of aspect ratios and consequently lead to interfacial area enhancement, a novel technique employing electron beam (e-beam) lithography coupled with dry etching strategies is presented [22].

4.2 Experimental Performance Evaluation of O_2 Etched Nafion Membranes

The performance of O_2 plasma etched Nafion with sputtered thin-film catalyst layer was examined. The blanket (i.e., non-area selective) etching of Nafion in O_2 plasma leads to the roughening of the membrane's surface and increased superficial area. The etching of Nafion was done in a Trion Phantom III RIE reactor. The flowrate, power, pressure, and duration were maintained at 40 sccm, 80 W RIE power, 6.7 Pa, and 60 s respectively. The etching process leads to the removal of approximately 1 μm of Nafion from the membrane's surface. Pt was sputtered onto both sides of the etched membrane from a Pt target using a Lesker PVD-75 sputter coater equipped with a crystal thickness monitor (Lesker gold electrode 6 MHz) for rate and layer thickness determination. The nominal sputter base pressure, gas pressure, power, and deposition rate were 8.0×10^{-4} Pa, 6.7×10^{-1} Pa, 75 W, and 0.1 nm/s respectively. The active area for the deposited electrodes was 1 cm^2. Pt was similarly sputtered onto an unetched Nafion sample for comparison. The Pt loading for the sputtered cells was 0.02 mg/cm^2

The sputtered catalyst coated membranes were assembled into test cells by sandwiching them between two carbon fiber paper (AvCarb™ GDS2120) GDLs. The formed MEAs were neither hot pressed nor was any ionomer solution added as binder. High purity H_2 and O_2 were used as fuel and oxidant respectively. The test cells were conditioned for approximately 24 h at 0.6 V, 60 °C cell temperature, 100 % RH, and 1 atm back-pressure. For performance testing, the anode humidifier, cell, and cathode humidifier were set to 58, 60, and 56 °C respectively, and the polarization data was obtained by gradually increasing the current produced from the PEMFC and recording the potential response. Linear voltammetry was performed on the test cells using a PAR 2273 system. H_2 and N_2 were fed as anode and cathode gases respectively. The cathode was used as the working electrode, while the anode was both the reference and counter electrode. For data collection, the potential of the working electrode was swept in the anodic direction from 0 to 0.6 V at a scan rate of 2 mV/s and the current response recorded. The limiting crossover current was obtained from the linear voltammetry experiment.

Polarization data was collected for the unetched, etched, and a 0.1 mg/cm^2 commercial MEA (Fuelcellstore.com). The commercial MEA possesses the

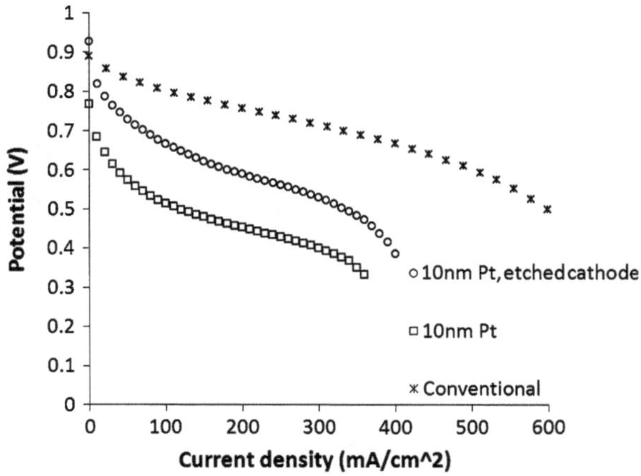

Fig. 4.3 Polarization performance curves obtained for conventional and sputtered catalyst cells

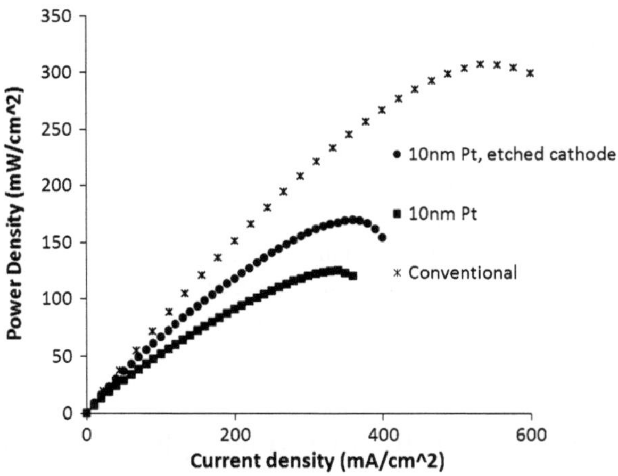

Fig. 4.4 Power density curves obtained for conventional and sputtered catalyst cells

conventional ink-based Pt supported-on-carbon catalyst layer. The result of this polarization test is shown in Fig. 4.3.

The sputtered cells show higher activation polarization losses compared to the commercial MEA. This was expected due to the much lower loading of the sputtered cells. In comparison with the unetched Nafion, the etched Nafion substrate had a lower activation loss. The sputtered cells also show a near parallel behavior around the ohmic polarization region. The limiting current densities obtained from the linear voltammetry scans for the unetched and etched sputtered cells were 0.23 and 0.30 mA/cm^2 respectively. Despite a larger crossover, the open circuit voltage of the etched membrane was higher than the unetched membrane. The etched membrane was expected to have a higher limiting crossover current density given that etching results in the thinning of the membrane, which reduces resistance for H_2 permeation and leads to a higher limiting crossover current. The power densities of the tested MEAs were similarly plotted in Fig. 4.4. The figure shows that the commercial MEA achieved the highest power density.

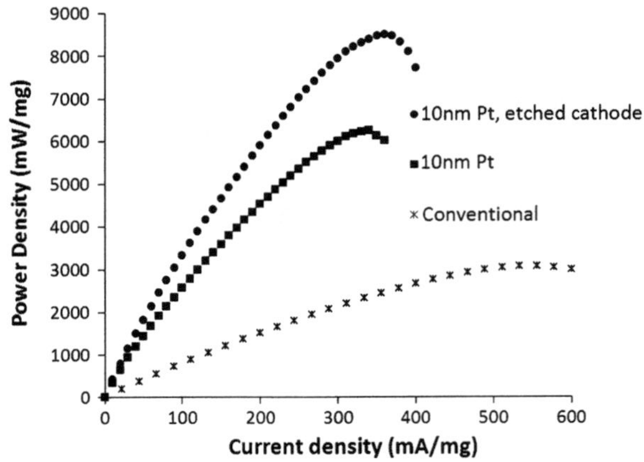

Fig. 4.5 Mass specific power density curves obtained for conventional and sputtered catalyst cells

The maximum power density of the commercial cell was approximately double and triple the etched sputtered and the unetched sputtered cells respectively. The mass specific power density, which is determined by dividing the area specific power density by the catalyst loading, is plotted in Fig. 4.5. The sputtered cells had higher mass specific power density compared to the conventional catalyst MEA, and with the O_2 etched membrane showing the best results.

Converse to the area specific power density, the sputtered cells show dramatically positive performance on a weight basis. Figures 4.4 and 4.5 together indicate the need for continued research on the sputtered cells toward increasing area specific power density without sacrificing mass activity.

4.3 Electron Beam Lithography for Nafion Patterning

Electron beam lithography (EBL) circumvents the current resolution limits of optical lithography, while dry etching schemes provide the ability to modify surface morphology and vary isotropicity during pattern transfer [4, 10, 19]. For patterning large areas, EBL can be used to fabricate optical or nanoimprinting masters after the identification of profiles that result in device performance enhancement via direct writing. Successful pattern transfer in EBL is a function of process parameters from the separate steps of EBL, masking, and etching. EBL involves the exposure of an electron sensitive material to a beam of electrons. The electrons interact with material making it easier to resolve the exposed part. The resulting profile from the exposure step is dependent on resist and masking layer thickness, accelerating voltage, exposure dose, and equipment calibration. Etching can be used for post development pattern transfer. Figure 4.6 shows the fabrication route for patterning a Nafion membrane.

Nafion NRE212 was mounted on a silicon wafer using Kapton tape. A thin layer of Ge ($\sim$30–60 nm) was sputtered onto the Nafion substrate to serve as a hard mask.

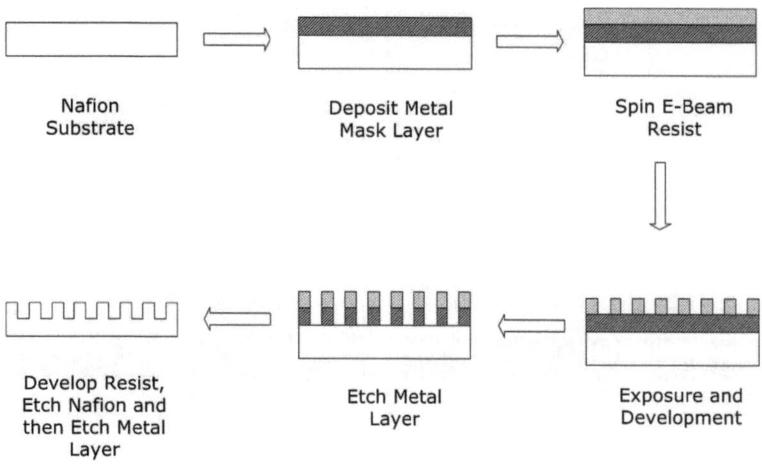

Fig. 4.6 Schematic diagram of EBL nanopatterning process

This was followed by spinning electron beam resist at 3 krpm for 45 s using a Brewster spinner and then baking at 70–75 °C. Zep520A, a positive resist, was used as the electron beam resist. Ge was selected as a masking layer because of its etch selectivity to both Nafion and the electron beam resist. The resist-coated Nafion substrate was exposed in a 100 kV JEOL JBX6300 EBL system. The system had a maximum writing field of 1,000 μm × 1,000 μm, pixel spacing as low as 1 nm, deflector speed of 50 MHz, and field stitching accuracy of 30 nm. For exposure, the beam current was typically set between 1 and 10 nA depending on required feature size and exposure time. A dose matrix test was used to determine the proper dosage for exposure. Generally, a reduction in beam dosage reduces exposure time. While this is beneficial for large area patterning, it may lead to compromised features with poor contrast. Nominal dosage was between 100 and 200 $\mu C/cm^2$.

After the exposure, the substrate was developed in xylene, but the subsequent rinsing in isopropyl alcohol (IPA) was not done, as Nafion immediately wrinkled when inserted in IPA. Dry nitrogen was used instead of IPA to drive off xylene after development. CHF_3 dry etching in a Trion Phantom III RIE reactor was employed to transfer the developed pattern into the hard mask layer, followed by cryogenic O_2 plasma etching in a Plasmalab100 ICP DRIE reactor to transfer the profile into Nafion and remove the organic resist layer. Both reactors permit control of RF power, pressure, etch time, and gas flow, but the Oxford Plasmalab reactor has additional and separate control of RF power for substrate bias. Figure 4.7 shows Nafion substrates patterned by electron beam lithographic and post-exposure dry etching steps. The cross-sectional images were obtained by sectioning and imaging the substrate with a Helios Nanolab SEM/FIB system.

The patterning process involves etching with O_2 and CHF_3. The dependence of the etch rate of Nafion on RF power was quantified for O_2 and CHF_3 plasmas. The RF power was varied from 80 to 200 W. The flow rate and pressure were maintained

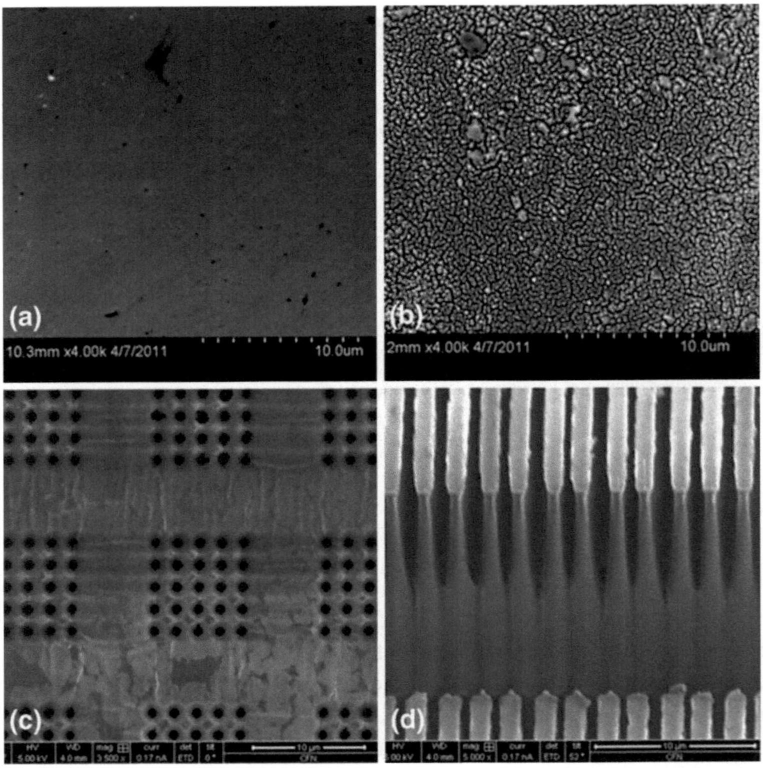

Fig. 4.7 SEM micrographs of **a** unetched, **b** O_2 plasma blanket etched, **c** EBL patterned dots, and **d** EBL patterned line structure Nafion membranes

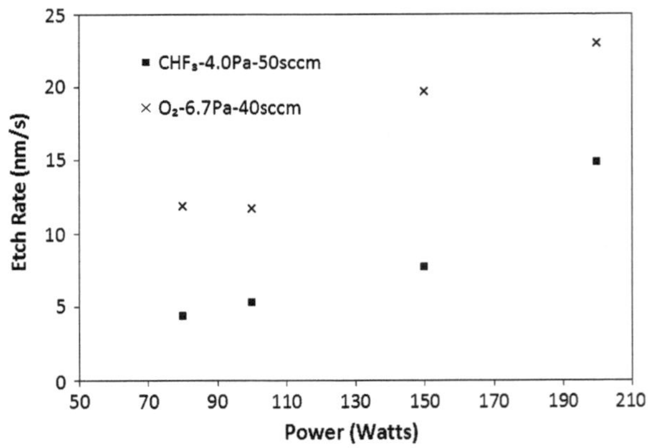

Fig. 4.8 Effect of RF power on Nafion etching in O_2 and CHF_3 gas plasmas

at 40 sccm and 6.7 Pa respectively for O_2 and 50 sccm and 4.0 Pa for CHF_3. The effect of power on etch rate of Nafion for both gases is shown in Fig. 4.8.

In Fig. 4.8, the etch rate of Nafion is seen to increase with power for both gases. An increase in RF power typically implies increases in both plasma density and ion

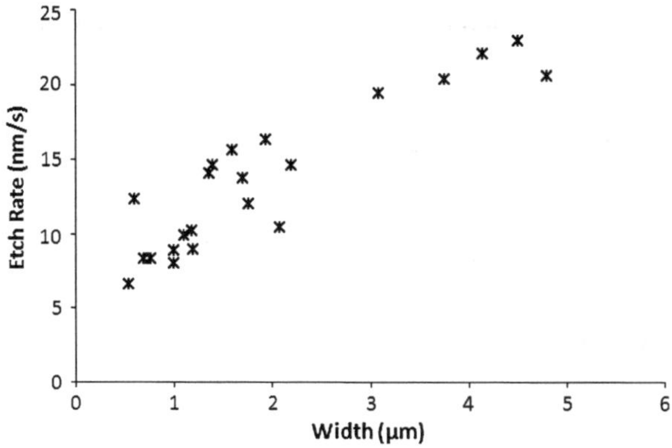

Fig. 4.9 Dependence of Nafion etch rate on critical dimension

energy [25]. As a result, higher fluxes of higher energy ions strike the substrate leading to increased physical etching. A higher plasma density can also result in enhanced chemical etching because of the production of more radicals and energetic reactive species that react with substrate material to form volatile products. Therefore, greater net etch rates can occur by increases in both physical and chemical components. Typically, a synergistic combination of both etching components is desired in order to optimize etch rate, anisotropicity, and possible substrate damage.

The dependence of etch rate on the feature size was explored. Figure 4.9 shows that etch rate diminishes with feature size. The figure also shows a bimodal variation of etch rate with cross-sectional width signified by the slope transition between 2 and 3 μm. These resulting variations of etch rate with critical dimension can be attributed to the relative ease with which (1) the active species can access the developing profile, and (2) the resulting products can exit. Higher aspect ratio results in mass transport limitations becoming increasingly dominant, leading to reduced etch rate through the limitations imposed by material transport and is representative of aspect ratio-dependent etching.

EBL coupled with dry etching strategies provide the ability to fabricate reproducible micro/nanoscale, high aspect ratio features on Nafion onto which catalyst layers can be deposited. It is envisioned that with optimization, this would lead to the fabrication of new PEMFC devices capable of providing high power densities at significantly low catalyst loadings.

4.4 Conclusions

This chapter introduced nano/micro fabrication innovations applied to the PEMFC toward the deployments of a cheaper, more efficient, and better performance device. Compared to battery systems, the PEMFC can be operated for much longer times making it preferable for continuously powering autonomous devices. Currently applied nanofabrication techniques including sputter deposition of catalyst layers, physical abrasion, sputtering, molding, and NIL techniques for patterning the Nafion membrane are tools which can enable performance enhancement of PEMFCs of various scales. A new technique leveraging the tools of EBL and advanced dry etching to produce high aspect ratio features of submicrometer critical dimension and larger is also discussed. This new capability opens the possibility of creating a variety of novel advanced architectures for PEM fuel cells with tunable properties.

References

1. Aizawa M, Gyoten H, Salah A, Liu X (2010) Pillar structured membranes for suppressing cathodic concentration overvoltage in PEMFCs at elevated temperature/low relative humidity. J Electrochem Soc 157(12):B1844–B1851
2. Banerjee S, Curtin DE (2004) Nafion® perfluorinated membranes in fuel cells. J Fluorine Chem 125(8):1211–1216
3. Barbir F, Yazici S (2008) Status and development of PEM fuel cell technology. Int J Energy Res 32(5):369–378
4. Campbell SA (2001) The science and engineering of microelectronic fabrication. Oxford University Press, New York
5. Challa VR, Prasad MG, Fisher FT (2011) Towards an autonomous self-tuning vibration energy harvesting device for wireless sensor network applications. Smart Mater Struct 20:025004
6. Charles C, Ramdutt D, Brault P, Caillard A, Bulla D, Boswell R, Rabat H, Dicks A (2007) Low energy plasma treatment of a proton exchange membrane used for low temperature fuel cells. Plasma Phys Control Fusion 49(5A):A73–A79
7. Chen D, Li W, Peng H (2008) An experimental study and model validation of a membrane humidifier for PEM fuel cell humidification control. J Power Sour 180(1):461–467
8. Cho YH, Bae JW, Cho YH, Lim JW, Ahn M, Yoon WS, Kwon NH, Jho JY, Sung YE (2010) Performance enhancement of membrane electrode assemblies with plasma etched polymer electrolyte membrane in PEM fuel cell. Int J Hydrogen Energy 35(19):10452–10456
9. De Almeida SH, Kawano Y (1999) Thermal behavior of Nafion membranes. J Therm Anal Calorim 58(3):569–577
10. Franssila S (2010) Introduction to microfabrication. Wiley, West Sussex
11. Gasda MD, Teki R, Lu TM, Koratkar N, Eisman GA, Gall D (2009) Sputter-deposited Pt PEM fuel cell electrodes: particles vs layers. J Electrochem Soc 156(5):B614–B619
12. Gasteiger HA, Markovic NM (2009) Just a dream or future reality? Science 324(5923):48–49
13. Grot W (2007) Flourinated ionomers (plastic design library fluorocarbon). William Andrew, New York
14. Gruber D, Müller J (2007) Enhancing PEM fuel cell performance by introducing additional thin layers to sputter-deposited Pt catalysts. J Power Sour 171(2):294–301
15. Heitner-Wirguin C (1996) Recent advances in perfluorinated ionomer membranes: structure, properties and applications. J Membr Sci 120(1):1–33
16. Joh HI, Ha TJ, Hwang SY, Kim JH, Chae SH, Cho JH, Prabhuram J, Kim SK, Lim TH, Cho BK, Oh JH, Moon SH, Ha HY (2010) A direct methanol fuel cell system to power a humanoid robot. J Power Sour 195(1):293–298
17. Kim SJ, Lee IT, Kim YH (2007) Performance enhancement of IPMC actuator by plasma surface treatment. Smart Mater Struct 16:N6
18. Lee SJJ, Cha SW, Hayre R (2001) Miniature fuel cells with non-planar interface by microfabrication. In: Jain M, Ryan A, Surampudi RA, Nagarajan G (eds) Power sources for the new millennium. The Electrochemical Society, Pennington, p 67

19. Madou MJ (2002) Fundamentals of microfabrication: the science of miniaturization. CRC, Boca Raton
20. Mauritz KA, Moore RB (2004) State of understanding of Nafion. Chem Rev 104(10):4535–4586
21. O'Hayre R, Lee SJ, Cha SW, Prinz FB (2002) A sharp peak in the performance of sputtered platinum fuel cells at ultra-low platinum loading. J Power Sour 109(2):483–493
22. Omosebi A, Besser RS (2011) Electron beam assisted patterning and dry etching of Nafion membranes. J Electrochem Soc 158(10):D603–D610
23. Ozturk O, Ozdemir OK, Ulusoy II, Ahsen AS, Slavcheva E (2010) Effect of Ti sublayer on the ORR catalytic efficiency of dc magnetron sputtered thin Pt films. Int J Hydrogen Energy 35(10):4466–4473
24. Pichonat T, Gauthier-Manuel B (2007) Recent developments in MEMS-based miniature fuel cells. Microsyst Technol 13(11):1671–1678
25. Plummer JD, Deal MD, Griffin PB (2000) Silicon VLSI technology: fundamentals, practice and modeling. Prentice Hall, Upper Saddle River
26. Prasanna M, Cho EA, Kim HJ, Lim TH, Oh IH, Hong SA (2006) Effects of platinum loading on performance of proton-exchange membrane fuel cells using surface-modified Nafion® membranes. J Power Sour 160(1):90–96
27. Shah K, Shin WC, Besser RS (2004) A PDMS micro proton exchange membrane fuel cell by conventional and non-conventional microfabrication techniques. Sens Actuators B Chem 97(2–3):157–167
28. Song C (2002) Fuel processing for low-temperature and high-temperature fuel cells: challenges, and opportunities for sustainable development in the 21st century. Catal Today 77(1–2):17–49
29. Yeom S-W, Il-Kwon O (2009) A biomimetic jellyfish robot based on ionic polymer metal composite actuators. Smart Mater Struct 18(8):085002
30. Taylor AD, Lucas BD, Guo LJ, Thompson LT (2007) Nanoimprinted electrodes for micro-fuel cell applications. J Power Sour 171(1):218–223
31. Wilhelm AN, Surgenor BW, Pharoah JG (2006) Design and evaluation of a micro-fuel-cell-based power system for a mobile robot. IEEE/ASME Trans Mechatron 11(4):471–475
32. Zhang Y, Lu J, Shimano S, Zhou H, Maeda R (2007) Nanoimprint of proton exchange membrane for MEMS-based fuel cell application. In: 6th International IEEE conference on polymers and adhesives in microelectronics and photonics, polytronic 2007, proceedings 2007, pp 91–95
33. Zhang Y, Lu J, Zhou H, Itoh T, Maeda R (2008) Application of nanoimprint technology in MEMS-based micro direct-methanol fuel cell (μ-DMFC). J Microelectromech Syst 17(4):1020–1028
34. Zhou Z, Dominey RN, Rolland JP, Maynor BW, Pandya AA, DeSimone JM (2006) Molded, high surface area polymer electrolyte membranes from cured liquid precursors. J Am Chem Soc 128(39):12963–12972

Chapter 5
Vibration Energy Harvesting and Its Application for Nano- and Microrobotics

Junjun Ding, Vinod R. Challa, M. G. Prasad and Frank T. Fisher

Abstract In this chapter, the concept of vibration energy harvesting and its potential utilization as a power source for micro/nanorobots is introduced and discussed. While batteries are commonly used as power sources for electronic devices, the limited lifetime and relatively large dimensional structure of batteries constrain its applications in micro/nanodevices. On the other hand, while capacitors have an extremely compact structure, the power stored is generally considered too low to power micro/nanorobots for a sufficiently long period of time. Hence, energy harvesting approaches, either alone or in conjunction with more traditional power sources, are being investigated to provide sufficient power for micro/nanorobots over the design lifetime of the system. In general, several varieties of energy harvesting techniques and devices have been developed to transfer different energy sources that may be present in a particular environment into electrical power; for example, solar, thermal, mechanical vibration, and even wind energy can be used as the source for energy harvesting devices in appropriate applications. However, specifically considering micro/nanodevices for biomedical applications greatly restricts the potential energy sources that can be harvested for system power. In this case, mechanical (vibration) energy may serve as a useful environmental source for energy harvesting. The purpose of this chapter is to first introduce the reader to the general field of energy harvesting, after which the discussion will focus on mechanical vibration energy harvesting and other techniques with potentially greater application to biomedical nano/microrobotics. In particular, different vibration energy harvesting mechanisms, such as electromagnetic, electrostatic, and piezoelectric techniques, will then be presented. The chapter concludes with recent work being done in the field of nanotechnology to further extend these energy harvesting approaches to size scales compatible with nano/micro devices and systems.

J. Ding · V. R. Challa · M. G. Prasad · F. T. Fisher (✉)
Department of Mechanical Engineering, Stevens Institute of Technology,
Hoboken, NJ 07030, USA
e-mail: Frank.Fisher@stevens.edu

Y. Guo (ed.), *Selected Topics in Micro/Nano-Robotics for Biomedical Applications*,
DOI: 10.1007/978-1-4419-8411-1_5,

5.1 Introduction

In the most general terms, energy harvesting (also referred to as energy scavenging or power harvesting in the literature) describes a process whereby energy in a given environment is transferred into a more useful form of electrical energy. Energy is the amount of power consumed, expressed in watt-hours or kilowatt-hours, while power is often expressed in watts or kilowatts. Energy is equal to the power multiplied by the time of consumption. Thus, in specific terms energy harvesting is somewhat of a misnomer, as technically the area should be referred to as power harvesting (although following common convention the term energy harvesting will be used here). The sources of energy that can be harvested are practically limitless, and in the most general terms can include sources such as solar energy, thermal energy, wind energy, tidal and wave energy, and mechanical and vibration energy. Further, the scales on which energy harvesting can be pursued can vary drastically, generating anywhere from utility-level power levels in applications such as hydroelectric dams, wind (turbine) farms, and solar (or photovoltaic) applications to energy harvesting at much smaller scales, where micro- (or even nano-) watt levels of electrical energy that are generated are useful for a small-scale device or system. It is the latter level of energy that is of particular interest for micro/nanorobots and will be the focus of the present chapter. Despite the broad, general definition of energy harvesting given above, in many cases (including in this chapter), the term energy harvesting is used in the context of much smaller levels of power which are likely to be on the orders of milliwatts (mW) for most micro/nanorobots applications.

While the power available from mechanical vibrations in an environment is generally small, the power requirement of micro/nanorobots is low and therefore this approach has great potential in the powering of micro/nanorobotic devices and systems. For example, a particular application of interest is the powering of future bio-microsystems that consist of low-power transducers, low-power integrated circuits (IC), and passive sensors, specifically designed to consume low amounts of power to extend their battery life. In order to place the power needed for microscale robots in perspective, the power consumed by insects of similar size can be illustrative. Results compiled in the literature [1] show that the common fruit fly (*Drosophila virilis*) has a mass of 0.001 g and requires 0.03 mW of power, while insects approximately 1 g in mass require approximately 1–50 mW of power, resulting in specific power (power per unit mass) values on the order of tens of W/kg. On the other hand, implantable intraocular pressure microsensors with a power consumption of 4.2 nW have been powered by microbatteries [2], while an artificial mechanical white cell of microscopic size called a microbivore may consume up to 200 pW of continuous power while completely digesting trapped microbes [3]. Such implantable biosensors are generally restricted in terms of the space available for power sources, thus power scavenging approaches are an attractive means to extend the operational lifetime of these systems.

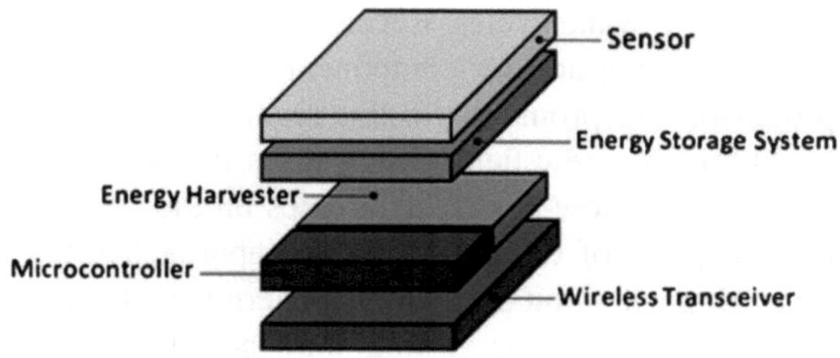

Fig. 5.1 Schematic of a wireless sensor node using a small-scale energy harvester to power data collection and wireless data transmission

Methods of storing and supplying power at small length scales is a challenging topic for micro/nanorobots, and likely represents one of the largest obstacles to realizing functional micro/nanorobots in the future. Because of the energy density and volume of batteries, they are generally not suitable (or promising) for micro/nanorobots. Recent advances seeking to leverage nanotechnology to further battery technology have met with mixed reviews, and while thin-film batteries may ultimately enable integration with MEMS devices, they are likely to struggle with providing a sufficiently long-term power source for micro/nanorobots. Thus, the projected limited lifetime of microbatteries has motivated the pursuit of energy harvesting approaches as a means to provide power to these systems.

In general, an energy harvesting approach requires a *transduction mechanism* which transfers energy from the ambient environmental source into electrical energy, and a *conditioning (and/or storage) system* which conditions the acquired electrical energy into a form which can be either immediately used by the larger system or appropriately stored for later use (Fig. 5.1). Here, the transduction mechanism is responsible for generating electrical energy from the ambient source energy, for example, mechanical vibrations or temperature gradients. Electrical circuitry is often needed to modify the harvested electrical energy, for example converting or rectifying an AC input into DC for storage in an appropriate electrical system, for example batteries or capacitors. Generally, the rate at which the energy is harvested is quite low, which requires a gradual filling of the electrical storage element until a threshold value has been achieved that indicates that sufficient energy is available for the desired operation. A common analogy to illustrate this principle is slowly filling a bucket with water, where the energy harvested and energy available in the energy storage system are represented by the water flowing into the bucket and the overall amount of water in the bucket, respectively.

As noted above this filling process is typically quite slow, and is generally accomplished while the system is in a low-power consumption mode (or sleep mode). Once a sufficient level of power is available in the system, the overall system 'wakes up' and expends energy in doing its programed task, after which the system re-enters its sleep mode and continues harvesting energy until sufficient energy is available for the next cycle to commence as governed by a microprocessor or alternative control mechanism. For example, as shown in Fig. 5.1, the harvester is used to ultimately power a sensor for data collection and a transmitter for data transmission. Clearly, a careful balance of the energy harvested to the energy required by the overall system will largely control the ratio of the length of the sleep-to-wake cycles. Recent advances in low-power electronics related to the power consumption for components such as the sensors, transmitters, and micro-processors are promising in this regard.

Although a new field, as advances in energy harvesting methodologies on the one hand and low-power electronics on the other converge it will greatly expand the prospects of energy harvesting approaches for small-scale robotics applications. Depending on the available energy and the power requirements of the micro/nanorobot, energy harvesting may be suitable for either a stand-alone power source, where the entire system is powered via the harvested energy, or alternatively used in trickle-charge mode, where energy harvesting is used to supplement (and hence extend the lifetime of) more traditional power sources such as batteries. A key factor from an overall system design standpoint will be that the extra cost and complexity associated with energy harvesting is outweighed by the enhanced system performance that energy harvesting adds to the system. In this sense, the low size and power requirements and inaccessibility of micro/nanorobots when used for biomedical applications seems to provide a suitable niche for continued research in this area.

5.2 Alternative Power Sources for Micro/Nanorobotics

5.2.1 Batteries

Because of their relatively high power density, portability, and interchangeability, batteries are the most commonly used power source in electronics ranging from sensors and cell phones to computer laptops and other consumer products. Batteries not only can be used as a power source, but also as a power storage device as in the case of rechargeable batteries. Batteries are in general simple to implement, can provide a continuous power supply, and can be readily replaced as necessary in many applications where they are accessible.

Today, lithium-ion batteries are rapidly being developed to replace nickel metal hydride (NiMH) and nickel cadmium batteries for portable and hand held devices. Lithium-ion batteries can achieve specific energy levels of 120–130 Wh/kg, which is more than 50 % higher than that of typical portable NiMH batteries [4]. Lithium-ion batteries have a relatively long cycle life, broad operating temperature range, low self-discharge rate, rapid charge capability, relatively high energy density, high voltage output, and a lack of memory effect compared with other types of electrochemical batteries [5]. Disadvantages of lithium-ion batteries include capacity/power degradation at high temperatures and critical safety issues. In particular, lithium-ion batteries are sensitive to overcharging, which can lead to overheating and internal short-circuiting which adversely affects the battery performance. The heat released from the lithium-ion batteries is significantly large and can be problematic in some applications [6].

The primary disadvantages of current batteries are the power-to-size and power-to-weight ratios, which have not kept pace with the rate of miniaturization of electronics and other components that might make up a nano/microrobot or system. This leads to the current situation where the relative weight of the battery, as a percentage of overall system weight, increases as the electronic device is reduced in size. For example, for a current generation laptop computer, the battery is 13 % of the total mass, while in a Motorola cell phone the battery comprises up to 36 % of the total mass [5]. Recently, microbatteries using thin- and thick-film technology have been developed to decrease the volumes of batteries and to facilitate integration with Complementary Metal Oxide Semiconductor (CMOS) IC and devices such as microprocessors [7]. Since the amount of active material in the cells is reduced, the thin-film cell microbatteries have low current output and capacity, and thus the lifetime of microbatteries is generally limited although researchers are working to address this limitation [8, 9]. In many respects, battery technology is a significant bottleneck to the realization of further reductions in the size and weight of portable electronics, as battery technology has not been able to keep up with the pace of miniaturization of other electronics components. This limitation is clearly of critical concern for future micro- and nanorobotics applications.

5.2.2 Fuel Cells

Like batteries, fuel cells generate power through an electrochemical process. In general, a fuel cell consists of an anode and cathode separated by an electrolyte as shown in Fig. 5.2. A chemical reaction takes place at the interfaces of these three elements such that as the fuel is consumed electrical current (and hence electrical energy) is generated and some byproducts produced. In a common hydrogen fuel cell, H_2 and O_2 are supplied to the anode and cathode, respectively, with the byproduct H_2O being emitted from the cathode side. For a direct methanol fuel cell, methanol and H_2O and O_2 are supplied to the anode and cathode, respectively, with CO_2 and H_2O byproducts generated at the anode and cathode, respectively. For example, the direct methanol fuel cell reaction equations are shown as following [10],

$$\text{Anode}\quad CH_3OH + H_2O \rightarrow 6H^+ + 6e^- + CO_2$$
$$\text{Cathode}\quad \tfrac{3}{2}O_2 + 6H^+ + 6e^- \rightarrow 3H_2O$$
$$\text{Overall reaction}\quad CH_3OH + \tfrac{3}{2}O_2 \rightarrow 2H_2O + CO_2$$

Large-scale fuel cells can achieve 40–70 % conversion efficiency [11]; meanwhile, fuel cells operating on pure hydrogen have no emissions but water. Hence, fuel cells can provide competitive energy densities with low emissions at larger length scales. However, due to the consumption of anode and electrolyte, the lifetime of fuel cells is limited to the availability to store the fuel. As with batteries, micro/nanoscale fuel cells generally have limitations with respect to power density

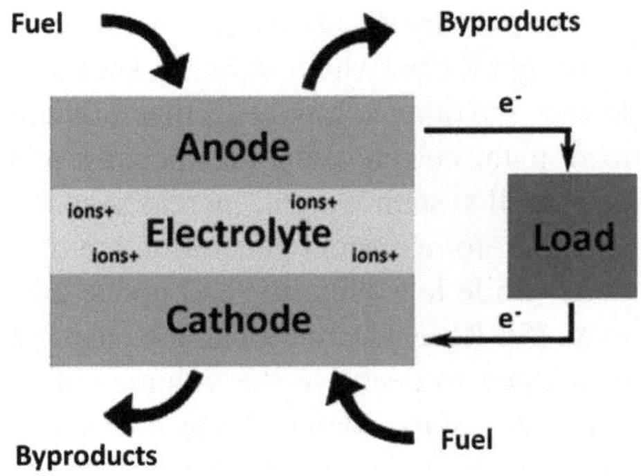

Fig. 5.2 Schematic of the operating principle of a fuel cell

and lifetime when implemented at smaller length scales [12]. Nonetheless, there is interesting work reported in the literature regarding the development of a particular subclass of glucose-based micro-fuel cells that can convert glucose readily available in the human body into electricity [13–16]. While outside the scope of this chapter, such results clearly warrant further study as possible power sources for micro/nanoscale robotics for biomedical applications. The potential of leveraging new advances in nanotechnology for low temperature fuel cells in microsystem applications is discussed in more detail in Chap. 4 of this book.

5.3 Energy Harvesting Mechanisms

Energy harvesting is being pursued as an alternative power source for small-scale electronic devices, largely driven by the desire to provide extremely long lifetime power sources for these systems. Energy sources ranging from mechanical motion and light radiation to temperature gradients can be used to generate electrical power. Solar-powered calculators, solar-powered battery rechargers, and photovoltaic solar cells have been widely used, whereas thermoelectric powered wristwatches driven by the temperature gradients between human skin and the environment are commercially available. While photovoltaic cells are clearly not practical for micro/nanoscale robotics for biomedical applications, some work has been done to investigate thermoelectric (or in more general terms, thermal energy) harvesters (see, for example, the discussion in [17]). More recently, the idea of using mechanical vibrations for the generation of electrical energy has gained widespread interest as discussed later in this chapter.

5.3.1 Thermoelectric Energy Harvesters Based on Temperature Gradients

In 1822, Seebeck discovered that a temperature gradient between two different types of conducting materials generates a potential difference. For example, if

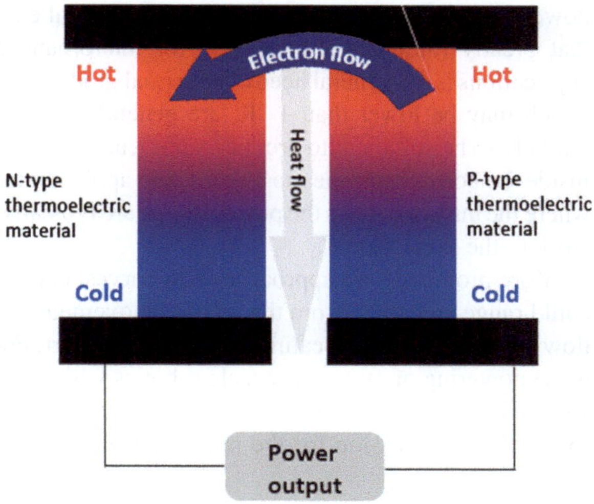

Fig. 5.3 Operating principle of a thermoelectric device

there is a temperature difference between the top and bottom surface as shown in Fig. 5.3, electron–hole pairs are created at the hot end due to the absorption of heat. The pairs then recombine and release heat at the cold end. The Seebeck voltage is created by the thermal difference. This potential voltage drives the flow of the electron–hole pairs, resulting in current being generated in the circuit [18].

The performance of thermoelectric devices mainly depends on the figure of merit (ZT) of the material, which expresses the efficiency of making electron–hole pairs and is given by [19]

$$ZT = \frac{\alpha^2}{\rho\lambda} T \tag{5.1}$$

where α is the Seebeck coefficient, ρ is the electrical resistivity, λ is the thermal conductivity, and T is the average temperature.

The conversion efficiency of thermoelectric generators is usually lower than 10 % [18]. From Eq. (5.1), in order to increase the conversion efficiency, the thermoelectric material should be a good electrical conductor (to minimize the heat consumed when current is passed through the material) and a poor thermal conductor (to maintain the maximum temperature difference between the hot and cold ends of the thermoelectric). In addition, the Seebeck coefficient should be maximized to produce the highest ZT value. However, it is difficult to improve the figure of merit, as for most materials the Seebeck coefficient, electrical resistivity and thermal conductivity are generally interrelated with each other. For example, increasing the Seebeck coefficient usually leads to a decrease of electrical conductivity, while increases in electrical conductivity are generally related to increases in thermal conductivity as well.

Nonetheless, thermoelectric generators based on temperature gradients between junctions of two different materials have been used for energy harvesting. For example, both Seiko and Citizen have developed commercialized thermoelectrically powered wristwatches [20]. However, for thermal energy harvesters, one challenge that greatly limits its applicability for micro/nanoscale robotics for biomedical applications is the general need of a thermal gradient. Slight temperature differences, which may be lower than 1 °K, are generally found in the human body and are unlikely to be sufficient to produce sufficient power for micro/nanorobots operating inside the body. Thus, the most promising applications of thermal energy harvesters where the human body is the power source are typically limited to their use on or very close to the skin surface.

More promising are approaches that harvest energy from kinetic sources, which could range anywhere from the relative movement of knees and shoulders to blood flow in the body to the beating of the human heart. For applications where the goal is the powering of micro/nanoscale robotics within the human body, kinetic energy sources that are interior to the human body are of primary interest and will be the focus of the discussion for the remainder of this chapter.

5.3.2 Mechanical Vibration Energy Harvesters

Vibrations are sources of (kinetic) mechanical energy that exist in almost every environment. There are two types of vibrations, free vibrations and forced vibrations. Free vibrations are initiated by an external source but continue to vibrate without the presence of an external force at the natural frequency of the system. These vibrations decrease in magnitude over time due to damping effects that are present in any system. On the other hand, forced vibrations are constantly subjected to an external force or motion. In free vibrations, energy is exchanged with the environment through damping and the process of harvesting electrical energy from these mechanical vibrations necessarily results in the additional damping impacting the behavior of the system. Both free and forced vibrations are of interest from the perspective of energy harvesting, as the particular type of vibration to be considered for a given application is predominantly based on the types of vibration available. One can readily observe the explosion of interest in the field of vibration energy harvesting in the first decade of the twenty-first century by examining the number of journal articles indexed per year in a popular science citation database as shown in Fig. 5.4. Further, more recently MEMS-scale vibration energy harvesting has been pursued by a number of researchers due to the potential benefits of harvesting at this length scale and the prevalence of vibrations in almost all environments. A summary of recent experimental power outputs under different conditions for vibration energy harvesters is shown in Table 5.1.

Vibration energy harvesters generate electrical energy from mechanical vibrations in the environment. In general, electrical energy harvested from mechanical vibration energy can be classified as resonance or non-resonance based [32]. In the resonance-based approach, the device is designed to be in resonance with the frequency of the environmental vibration source to obtain maximum

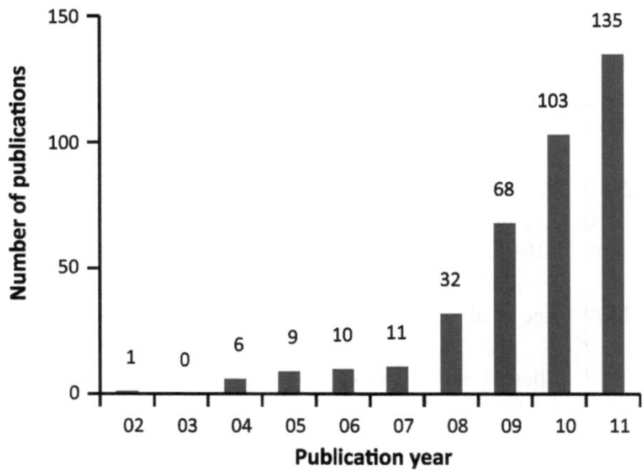

Fig. 5.4 Journal articles in the field of vibration energy harvesting since 2002. (*source*: Scopus, Keywords: 'vibration' AND 'energy harvesting')

deformation, displacement, or velocity of the transduction element in the energy harvesting device. In this case, maximum electrical energy will be generated by the transduction element. For example, engine vibrations from a car are suitable for such resonant generators. While the amplitudes of high frequency vibrations are generally small, resonance conditions ensure that maximum displacement of the transducer results in the maximum power output for a given vibration source. On the other hand, non-resonance energy harvesting devices are usually applied to lower frequency vibrations but require a suitable vibration source from which to harvest electrical energy (for example, structural element within a helicopter which undergoes consistent cyclic loading).

The major goal in designing an energy harvesting device is to achieve high efficiency and power density for a given mechanical vibration source. In order to design high efficiency and high energy density energy harvesting devices from mechanical vibration sources, appropriate models must be developed. The most commonly used model to describe kinetic energy harvesting systems was developed by Williams and Yates [33] and consists of the spring, mass and damper system shown in Fig. 5.5. Note that in Fig. 5.5 there are two separate dampers, labeled b_m and b_e, which represent the mechanical and electrical damping in the system, respectively. The electrical damping represents the electrical energy generated by the energy harvesting device.

In Fig. 5.5, it is assumed that the entire system is excited by an external vibration source $y(t) = Y_0 \sin(\omega t)$. It is further assumed that the mass of the vibration source is much greater than the mass of the energy harvesting device, such that the movement of the device will not affect the characteristics of the vibration source. In this case, the equation of motion will be

$$m\ddot{z}(t) + d\dot{z}(t) + k(z)t = -m\ddot{y}(t) \tag{5.2}$$

where m is the seismic mass, d is the damping constant, and k is the spring constant, and $z(t)$ is the relative displacement of the mass. The power transferred to the mass can be written as

Table 5.1 Summary of recent microscale vibration energy harvesting research

Years	Author	Mode of operation	Beam l × w (μm)	Source (m/s^2 @ Hz)	Power (μW)
2011	Massero et al. [21]	Piezo	600 × ~50 (ring-shaped due to stress)	19.6 @ 64	0.0013
2010	Park et al. [22]	Piezo	1000 × 800	3.8 @ 528	1.1
2010	Lee et al. [23]	Piezo	3000 × 1500	19.6 @ 206.8	1.8
2009	Elfrink et al. [24]	Piezo	1010 × 500	19.6 @ 572	60
2009	Lee et al. [25]	Piezo	1500 × 500 (31 mode) 1500 × 750 (33 mode)	9.8 @ 255.9 9.8 @ 214	2.765 1.288
2009	Shen et al. [26]	Piezo	3200 × 400	7.36 @ 184	0.32
2008	Shen et al. [27]	Piezo	4800 × 400	19.6 @ 465	2.15
2008	Kim et al. [28]	Piezo	1500 × 500	78.4 @ 870	1.13
2007	Marzecki et al. [29]	Piezo	1200 × 800	39.2 @ 1300	2
2007	Beeby et al. [30]	EM	9000 × 3000	0.59 @ 52	46
2006	Fang et al. [31]	Piezo	2000 × 600	9.8 @ 608	2.2
2005	Jeon et al. [32]	Piezo	260 × 170	106 @ 13.9 k	1

Piezo pieoelectric, *EM* electromagnetic

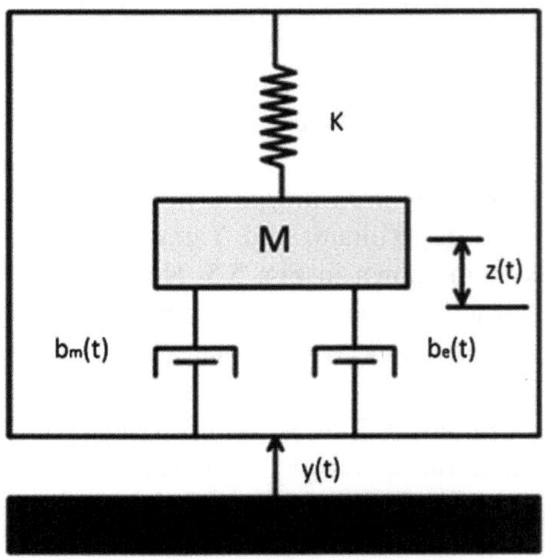

Fig. 5.5 Schematic of a generic energy harvesting device

$$P(t) = -m\ddot{y}(t)[\dot{y}(t) + \dot{z}(t)] \quad (5.3)$$

Further, it can be shown that the electrical power produced by the electrical damper can be written as [33]

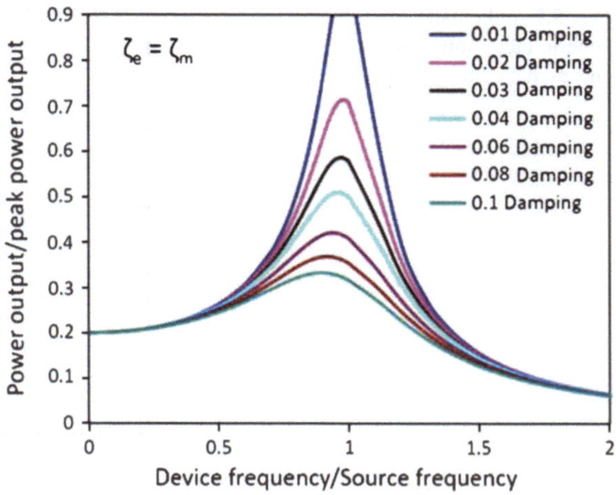

Fig. 5.6 Power generation versus source frequency for various damping ratios (normalized to peak power output at a damping of 0.01)

$$P_e(t) = \frac{m\zeta_t Y_o^2 \left(\frac{\omega}{\omega_n}\right)^3 \omega^3}{\left[1 - \left(\frac{\omega}{\omega_n}\right)^2\right]^2 + \left[2\zeta_t \frac{\omega}{\omega_n}\right]^2} \tag{5.4}$$

where Y_0 is the amplitude of vibration, ω is the angular frequency of the vibration source, ω_n is the natural frequency of the device, and ζ_t is the total damping ratio in the system (i.e. the sum of the individual damping components) such that

$$\zeta_t = \zeta_e + \zeta_m \tag{5.5}$$

where ζ_e is the electrical damping ratio and ζ_m is the mechanical damping ratio. Based on Eq. (5.4), the average power versus vibration source frequency for different damping ratios can be plotted as shown in Fig. 5.6. As expected, this plot shows that the power generated will be largest when the frequency of the environmental vibration source matches the natural frequency of the harvesting device.

When the energy harvesting device is operating in resonance, such that $\omega = \omega_n$, the energy harvested is inversely proportional to the total damping ratio ζ_t as shown via substitution into the Eq. (5.4),

$$P_{resonance} = \frac{mY_0^2\omega_n^3}{4\zeta_t} \tag{5.6}$$

As expected, Fig. 5.6 illustrates that the power generated at resonance under different damping ratios will be higher for smaller values of damping ratio. When the damping ratio $\zeta_t = 0$, the theoretical power generated will be infinite at resonance. However, due to inevitable damping such as friction and air resistance, this situation will not happen in practice. As the damping ratio decreases, the displacement of the seismic mass will increase; however, this displacement is usually limited by the size of the generator or device. Clearly, the system must be designed

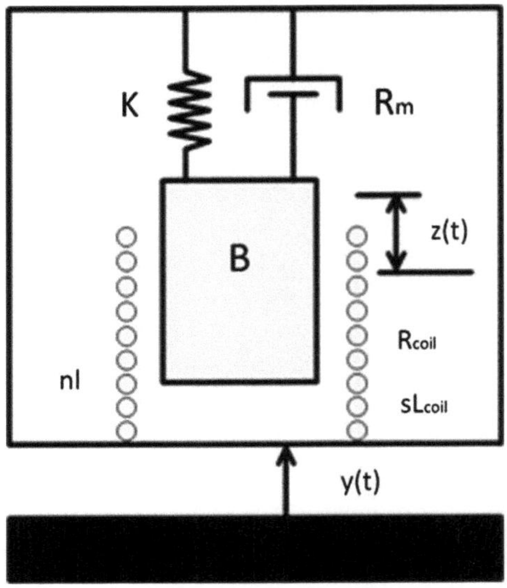

Fig. 5.7 Schematic of a generic electromagnetic energy harvesting device

to prevent the displacement of the mass from exceeding the maximum allowable displacement to protect the device from damage. Equation (5.6) suggests that the maximum power output will be obtained with the heaviest allowed seismic mass that does not result in failure of the vibrating system (such as material yielding, fatigue failure, etc.). Lastly, Eq. (5.6) shows that, for a constant excitation amplitude, power generation is proportional to the cubic of natural frequency, which indicates higher frequency vibration sources produce higher power.

The discussion above has been for the case of a generic energy harvester and is independent of the type of transduction mechanism employed in the device. Typically, vibration energy harvesting utilizes electromagnetic, electrostatic, or piezoelectric transduction approaches as discussed below.

5.3.2.1 Electromagnetic Transduction Mechanism

Electromagnetic vibration energy harvesting produces electricity from the motion of a magnetic material relative to a conductive coil. There are generally two ways to realize the relative motion between the magnetic material and the conductive coil, either moving the magnet or the coil. (Changing the magnetic field is an alternative way to produce current in the coil, although this is generally not practical from the perspective of energy harvesting). The electrical power generated will be a function of magnetic field, relative motion, and number of turns of the coil. A schematic of an electromagnetic energy harvesting device is shown in Fig. 5.7.

The electrical damping ratio can be determined by [5]

$$\zeta_e = \frac{(NlB)^2}{R_{\text{load}} + R_{\text{coil}} + j\omega L_{\text{coil}}} \tag{5.7}$$

where N is the number of turns in the coil, B is the magnetic flux density, l is the coil length per turn, R_{coil} is the resistance of coil, R_{load} is the resistance of the load, and L_{coil} is the coil inductance. By adjusting the load resistance such that the electrical damping ratio ζ_e is equal to the mechanical damping ratio ζ_m, maximum power can be obtained [5]. The maximum average power provided to the load resistance under resonance conditions can be written as [5]

$$P_{\mathrm{em-max}} = \frac{mA^2}{16\zeta_m \omega_n}\left(1 - \frac{R_{\mathrm{coil}}}{R_{\mathrm{load}}}\right) \tag{5.8}$$

where A is the excitation acceleration and ζ_m is the mechanical damping ratio.

Since a low voltage is generally obtained from the electromagnetic technique, a voltage up-converter is generally required for powering actual devices; the corresponding electrical losses must be accounted for in the calculation of total power harvested. Early work in this area includes the development of an electromagnetic device that employs two magnets and four magnets coupled to a coil attached to a cantilever beam [34]. An electromagnetic based generator for generating power from ambient vibrations has also been presented [35]. Microscale electromagnetic energy harvesting devices have also been pursued [36]. The reader is directed to comprehensive reviews available in the literature for more discussion of the electromagnetic approach to vibration energy harvesting [5, 37].

5.3.2.2 Electrostatic Transduction Mechanism

Electrostatic vibration energy harvesters produce electrical energy by employing two conductive plates moving relative to one another that are electrically isolated by air, vacuum, or dielectric insulator to form a capacitor. The change in capacitance can be achieved by a mechanical forcing function to generate electrical energy, either through altering the distance between the plates, the common area between the plates, or the dielectric material between the plates. The electrostatic energy harvesting technique offers a major advantage of having the potential for implementation via silicon micromachining processing allowing ease of integration with electronics.

There are two ways to convert energy from an electrostatic energy harvesting approach, referred to as either charge (of the capacitor) constrained or voltage constrained [38]. The voltage constrained mechanism generally provides a higher energy output than the charge constrained approach; however, it requires an additional voltage source which significantly complicates the device fabrication [39] (while any electrical losses due to the additional voltage source would need to be factored into efficiency calculations for the harvester). Thus, for simplification only the power generated from the device for a charge constrained model is discussed here, which for a source frequency ω_s can be written as [40]

$$P_{\mathrm{es}} = \frac{1}{2} V_{\mathrm{in}}^2 \omega_s (C_{\mathrm{max}} - C_{\mathrm{min}}) \left(\frac{C_{\mathrm{max}} + C_{\mathrm{par}}}{C_{\mathrm{min}} + C_{\mathrm{par}}}\right) \tag{5.9}$$

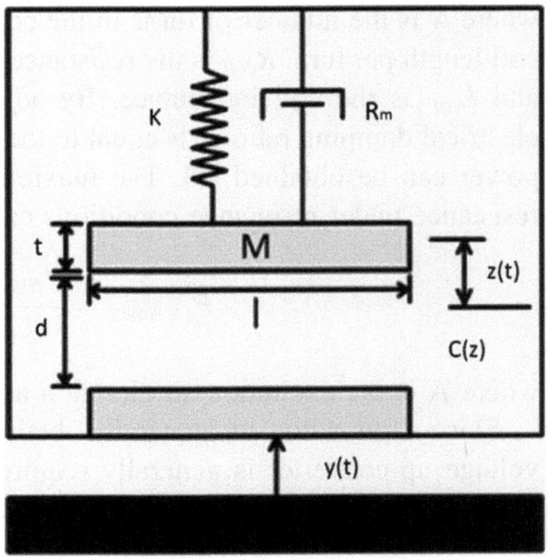

Fig. 5.8 Schematic of a generic electrostatic energy harvesting device

where $C_{\max}$, $C_{\min}$ and C_{par} are the maximum, minimum, and parasitic capacitance of the variable capacitor, respectively, and V_{in} is the input voltage of the capacitor.

To date, several examples of electrostatic vibration energy harvesting work at the microscale have been described in the literature. For example, Roundy et al. [41] and Chiu et al. [42] have made efforts to develop MEMS-scale devices that can produce power area densities of 116 μW cm^{-3} and 200 μW cm^{-2} at 120 Hz. Another work involves the development of a silicon-on-insulator (SOI) based MEMS electrostatic energy harvesting device consisting of comb structures and a seismic suspended mass [43]. Power scavenging of human motion is also being pursued, with one example being the powering of a cardiac pacemaker from the left ventricular wall motion of lab animals under development [44].

To improve the energy density of electrostatic generators, the capacitor spacing or surface area can be reduced. Although electrostatic generators are easily fabricated for MEMS applications, there are still several obstacles for commercial energy harvesting applications. Most importantly, the electrostatic devices require an input voltage (and hence input electrical energy) that may not be easily provided and in any event requires energy to produce the harvested energy. The output impedance and voltage of the electrostatic devices are often quite high resulting in very low current that may be ill suited for a particular application. Lastly, this limited amount of current often needs to be signal processed to be integrated within MEMS devices, which complicates the fabrication process.

5.3.2.3 Piezoelectric Transduction Mechanism

Piezoelectric energy harvesting is based on the piezoelectric property observed in some materials, which generates electrical charge when a mechanical stress is induced on the material. The electrical charge and voltage produced by piezoelectric materials depend on the mechanical, electrical and material properties of

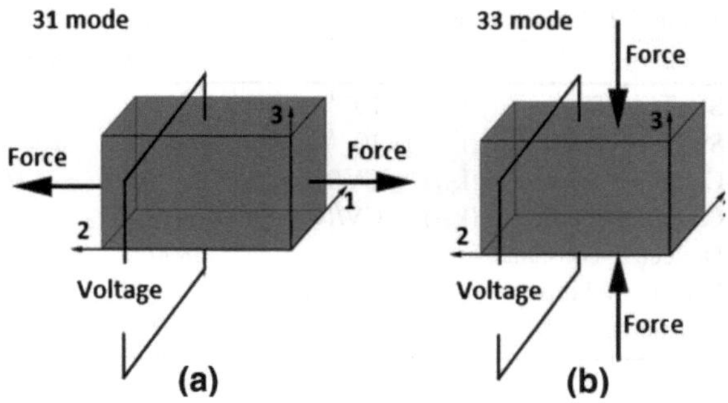

Fig. 5.9 **a** d_{31} mode of loading for piezoelectric structures **b** d_{33} mode of loading for piezoelectric structures

the piezoelectric materials and the stress applied on the material. Piezoelectric-based vibration energy harvesting is increasing in popularity due to the simplicity in capturing and storing the electrical energy generated; for example, a high voltage output enables relatively simple circuitry for capturing the power output, although significant gains can be realized by optimizing the circuitry [45]. Because electrical energy is produced simply due to the piezoelectric property of materials, such materials are typically employed as the vibrating structure. The constitutive equations of piezoelectric energy harvesting can be written as

$$\begin{bmatrix} S \\ D \end{bmatrix} = \begin{bmatrix} s & d \\ d & \varepsilon \end{bmatrix} \begin{bmatrix} T \\ E \end{bmatrix} \tag{5.10}$$

where S is the strain in the piezoelectric material, D is the electrical displacement, s is compliance, d is piezoelectric strain coefficient, ε is the dielectric, T is the stress, and E is the electric field.

Typically, energy is harvested in one of two piezoelectric modes (d_{31} and d_{33}) corresponding to the direction of mechanical stress and electrical output of the piezoelectric material/device. In d_{33} mode, the charge is collected on the electrode surface parallel to the stress direction, whereas in d_{31} mode the charge is collected on the electrode surface perpendicular to the direction of the mechanical force as shown in Fig. 5.9.

The coupling factor of d_{33} mode is generally higher than that of d_{31} mode resulting in higher energy conversion; however, the d_{31} mode enables larger strain to be produced by a low amplitude environmental vibration source and is generally more suitable at the microscale. The power output from the more common d_{31} mode piezoelectric energy harvester can be written as

$$P = \frac{\left(\frac{-d_{31} t_p \sigma_{\text{beam}}}{\varepsilon}\right)^2 R_L}{(R_S + R_L)^2} \tag{5.11}$$

where t_p is the thickness of the piezoelectric layer, σ_{beam} is the induced stress in the beam due to mechanical vibrations, ε is the dielectric constant of the piezoelectric material, and R_L and R_S are the load resistance and the magnitude of the impedance of

Table 5.2 Common piezoelectric materials and their representative bulk properties

Property	Units	PZT-5A	PZN-PT	ZnO	AlN	PVDF
Strain coefficient (d_{31})	10^{-12} m/V	−320	−950	−5.0	−2.6	10
Strain coefficient (d_{33})	10^{-12} m/V	650	2000	12.4	5.5	−38
Coupling coefficient (k_{31})	CV/Nm	0.43	0.5	0.196	0.23	0.12
Coupling coefficient (k_{33})	CV/Nm	0.75	0.91	0.48	–	0.15
Dielectric constant	$\varepsilon/\varepsilon_0$	3800	4500	10.9	12	10.46
Elastic modulus	10^{10} N/m^2	4.90	0.83	2.9	33	0.3
Tensile strength	10^7 N/m^2	2.0	8.3	4.0	–	5.1

the piezoelectric beam (source), respectively. As discussed previously, when the electrical load matches the impedance of the beam, a peak power output is obtained [5]. Representative material properties for common piezoelectric materials being pursued for energy harvesting applications are shown in Table 5.2; such values are approximations as the properties can be quite dependent on the exact composition and crystal structure of the processed material. Lead Zirconate Titanate (PZT) is widely used for piezoelectric applications due to its high coupling factor, although the presence of lead may be problematic for some applications. Newer materials, such as PZN-PT and PMN-PT, are also under development although not as commonly used in the energy harvesting community at this time [46]. On the other hand, Zinc Oxide (ZnO) and Aluminum Nitride (AlN) couple good piezoelectric properties with advantages such as ease of processing (particularly at the microscale) and their compatibility with a number of different environments. The piezoelectric polymer PVDF has also recently been used to demonstrate energy harvesting from low-speed air flow meant to mimic respiration [47].

Recent work has investigated the design of linear and nonlinear piezoelectric energy harvesters which can continuously recharge the batteries of human pacemakers by converting the vibrations from the heartbeats to electrical energy [48] using a zig-zag beam structure for the vibrating piezoelectric cantilevers [49, 50]. Constrained in a small volume, the frequency of a standard cantilever beam will be too high to be in resonance at heartbeat frequency, which is around 60 Hz. The zig-zag geometry can be designed to lower the natural frequency of the device to match that of the environmental source. By adding a nonlinear magnet force to the tip mass, the energy harvester becomes nonlinear with higher frequency bandwidth [48]. Of importance is that the nonlinear energy harvesting device can be designed to be relatively insensitive to the heart beat frequency, which varies from different people and at different times. In addition, piezoelectric energy harvesting for other implantable devices has also been investigated. For example, a square PZT membrane driven by a fluctuating pressure source (blood pressure) was used to produce power for bio-MEMS applications [51]. Although the devices successfully produced power, the amount of power produced at this time was insufficient for the proposed application. Energy harvesting from piezoelectric materials integrated in footwear has been designed for powering low-power wearable electronic devices [52], while the piezoelectric polymer PVDF has been applied in shoes to harvest energy from walking.

Further discussion of piezoelectric approaches to energy harvesting can be found in the excellent review articles of Beeby et al. [38], Sodano et al. [53], and Anton et al. [54], while Dutoit et al. [33] specifically discuss the design considerations for developing MEMS-scale piezoelectric energy harvesting devices.

In summary, vibration-based energy harvesting devices have potential applications in wireless sensor networks, condition monitoring, wearable self-powered intelligent devices, and medical implants areas. Electromagnetic and piezoelectric transducers have been widely investigated, whereas electrostatic transducers are less commonly pursued for practical applications. Piezoelectric transducers are promising for their simple structures and integration in MEMS using thin- or thick-film fabrication techniques, while electromagnetic transducers are not as easily integrated in MEMS devices due to the challenge of integrating the magnet and conductive coil within the device.

5.4 Nanofibers for Energy Harvesting Applications

It would be remiss in a chapter discussing vibration energy harvesting and its application to nano- and microrobotics to not highlight some of the very exciting research being described in the literature with regards to the fabrication and characterization of multifunctional nanostructures. Nanowires are one-dimensional materials with diameters of typically tens of nanometers and several microns (or more) in length. Many different types of nanowires have been developed, including metallic (Ni, Pt, Au), semiconducting (Si, InP, GaN), and insulating (SiO_2, TiO_2) nanowires. Of particular interest here is the development of nanowires based on materials that, at larger length scales, possess various properties that could be useful from an energy harvesting perspective such as Zinc Oxide (ZnO) and Lead Zirconium Titanate (PZT). (See, for example, the PZT nanowires described in Chap. 2 of this book.) Here, we briefly highlight some of the recent literature describing the development of nanofibers and nanofiber-based generators for energy harvesting applications.

5.4.1 ZnO-Based Nanofibers and Nanogenerators

The piezoelectric potential of ZnO nanowires is created by the polarization of ions in the crystal when the ZnO nanowires are elastically deformed. Unlike free mobile charges, the charges of the ions are fixed with atoms and cannot freely move. Thus, these charges cannot be completely removed by the free carriers in the semiconductor nanowires. For a nanowire of 50 nm in diameter and 500 nm in length, *if* there is no conductivity of the nanowire, one can show that the piezoelectric potential will be around 0.6 V. Considering the conductivity of ZnO with a carrier density of 10^{16}–10^{17} cm^{-3}, the magnitude of the piezoelectric potential will be around 0.3 V, which is sufficient to operate a Schottky diode [55]. Experimental results of the piezoelectric potential were measured directly by a micrometer probe compressing the side surface of a ZnO nanowire [56].

ZnO nanowires provide a number of advantages with respect to the development of a mechanical vibration energy harvesting device [57]. ZnO nanowires can be subjected to extremely large elastic deformation without fatigue or failure. At the same time, nanowires are typically free of dislocations and thus have a high resistance to fatigue due to the small diameter, which would tend to extend the lifetime of ZnO nanowire-based generators. Extremely weak mechanical disturbances can be collected by the nanowires to produce electrical output, which increases the types of energy sources that can be used. In addition, ZnO is a biocompatible material that is suitable for applications in biological or medical areas, such as powering micro/nanorobots in human body. Nanowires also provide a unique opportunity for functionalization to improve the physical or chemical properties of the surface. Lastly, ZnO nanowires can be applied to sensors and transducers because of its piezoelectric and semiconducting properties.

Piezoelectric nanogenerators based on ZnO nanowire arrays were first developed by Wang et al. in 2006 [56]. The efficiency of the ZnO nanowire-based generator was estimated to be 17–30 %. The mechanism of nanowire-based generator is to couple piezoelectric and semiconducting properties of ZnO as well as the formation of a Schottky barrier between the metal and ZnO contacts. By leveraging both piezoelectric and semiconducting properties of ZnO nanowires, ZnO nanogenerators can be designed to convert mechanical energy into electrical energy. As highlighted below, two different structures of the ZnO-based nanowire nanogenerators have been pursued: vertical nanowire and lateral nanowire integrated nanogenerators.

The lateral arrangement of ZnO nanowires appears to date to be the most common structure. For example, a single ZnO nanowire generator has been formed by bonding both ends of the nanowire with metal contacts on a flexible polymer substrate [58], while as one might expect arranging an array of ZnO nanowire (also on a plastic substrate) resulted in a nanogenerator with larger power output [59]. For the latter example, vertically aligned ZnO nanowires were first transferred to the plastic substrate to form the laterally arranged nanowires, with parallel electrode arrays then deposited to connect the nanowires. The nanogenerator was found to produce an open-circuit voltage up to 2.03 V and a peak output power density up to 11 mW/cm^3 [59]. The output power is high enough to light up a commercial LED bulb, which shows the potential practical application of the nanogenerator based on ZnO nanowires.

On the other hand, the vertical integration of ZnO nanowires provides another approach to the development of nanogenerators. For example, nine vertically aligned individual ZnO nanowires were integrated within a nanogenerator which was found to exhibit a peak output voltage of 0.045 V and current of 2.5 nA within this single layer structure [60]. A vertically integrated multilayer nanogenerator can produce an output power density of 2.8 nW/cm^2 with a peak output voltage of 0.15 V and current of 7.2 nA [60]. Integration of individual ZnO nanogenerators into three-dimensional structures has been suggested as a means to improve the electrical power density of the nanogenerator. While in both cases these exciting preliminary results are promising, significant research challenges remain in moving this technology toward future applications in the nano/microrobotics area.

5.4.2 *PZT-Based Nanofibers and Nanogenerators*

Lead Zirconate Titanate [$PbZr_{1-x}Ti_xO_3$ (PZT)] is a well-known piezoelectric material with a high electromechanical coupling coefficient, high dielectric constant, and large remnant polarization [61]. PZT exhibits a much higher dielectric constant and piezoelectric strain coefficient than ZnO, which makes PZT a promising material for energy harvesting as shown earlier in Table 5.2. However, from recent developments in the fabrication of PZT nanofibers it has been shown that such nanofibers may have an even higher piezoelectric voltage constant, greater bending flexibility, and higher mechanical strength than bulk PZT which would enable its use in nanogenerators [62]. Given these properties PZT nanofibers would likely generate higher voltages and power outputs than other types of piezoelectric materials (for a given volume and same energy source) [63].

The properties of PZT nanofibers have generated significant interest in taking advantage of these nanofibers in the development of nanogenerators. In one such work, a nanogenerator based on laterally aligned electrospun PZT nanofibers on interdigitated electrodes was demonstrated [62]. Here, the PZT nanofibers had diameters of approximately 60 nm and lengths of 500 μm (these dimensions can to some extent be controlled by varying the fabrication parameters used in the electrospinning process). The final device is covered with PDMS polymer to protect the underlying nanowires and electrode structure. In the demonstration of the device, an alternating pressure applied on the top surface of the nanogenerator is transferred to the PZT nanofibers, such that the resulting tensile and bending stresses generate a voltage in the nanofibers. Here, an integrated electrode array is used to enhance the output voltage and power by combining each cell in parallel. The PZT fiber-based nanogenerator was found to experimentally produce a peak output voltage of 1.63 V and output power of 0.03 μW with a load resistance of 6 MΩ [62]. While significant work must be done to continue the development of these nanowire-based generators toward applications in nano- and microrobotics, the initial results demonstrated to date for both ZnO and PZT nanowires is indeed quite promising.

5.4.3 Other Types of Nanofibers with Applications in Energy Harvesting

In addition to vibration energy harvesting, the development of multifunctional nanowires is being pursued in the area of thermoelectric energy harvesting. In practical application, the super-thin structure of a nano/microscale thermoelectric generator could be integrated onto silicon chips to harvest heat energy produced in electrical circuits. The potential ease of integration of a silicon nanowire-based thermoelectric generator offers the prospective of an alternative way of energy harvesting. Although the output voltage of such a device would be relatively low, there may be potential applications in ultra-low, self-powered MEMS devices.

For example, a high-density silicon nanowire-based thermoelectric generator prepared by a top-down CMOS-compatible technique has been reported in the literature [64]. The fabricated thermoelectric generator can produce an open-circuit voltage of 1.5 mV and a short-circuit current of 3.79 μA under an estimated temperature gradient of 0.12 K. This result is of particular interest since bulk silicon had not previously been strongly considered for thermoelectric applications prior to this work; however, the efficiency of the thermoelectric modules of silicon nanowires is greatly improved due to the phonon boundary scattering at nanoscale [64]. The processing of this silicon nanowire-based thermoelectric generator is based on typical CMOS-compatible top-down processes, which is a very promising aspect of this approach. However, thermoelectric generators are based on temperature gradients (and not absolute temperatures), and significant temperature gradients are not generally available within the human body. Thus, the application of such thermoelectric nanogenerators for nano/microrobotics for biomedical applications may be limited.

In summary, nanowire-based nanogenerators provide a potential method to transfer mechanical energy in a given environment to useful levels of electrical energy necessary to power micro/nanodevices. Both ZnO and PZT nanowires have been used, and the results demonstrated to date are a promising first step toward the development of this technology for future practical applications. Due to the biodegradable and biocompatible properties of ZnO, generators based on ZnO nanowires have a special advantage in the medical or biological arena, whereas PZT nanofibers, which have much higher dielectric constant and piezoelectric voltage constant, can generally produce more electrical energy at this time. Nanowire technology is also being pursued to develop other energy harvesting approaches, such as their use for thermoelectric energy harvesting highlighted briefly here. While such work is clearly at the early stages of research, the results to date are quite intriguing and suggest that the idea of self-powered nano/microdevices using such nanowires may be possible in the future.

5.5 Conclusions and Outlook

Energy harvesting refers to the process of converting energy in a given environment into a more useful form of electrical energy that can be used by the system. Because the amounts of energy which can typically be extracted in this regard are quite small, a primary challenge in the energy harvesting community is identifying potential applications where the amounts of energy harvested, while small, are attractive such that the benefits of an energy harvesting methodology offset the extra costs and complexity of adopting this approach. As discussed in this chapter, the power consumed by small insects can be illustrative in identifying the levels of energy that would be useful for nano/microrobotic applications. In particular, because the miniaturization of batteries and other electrical energy storage elements has not kept pace with the advances in other areas of microsystem development, there is a concern that battery technology may represent a significant bottleneck to the realization of further reductions in the size and weight of portable electronics. This limitation is clearly of critical concern for future micro- and nanorobotics applications. Nano- and microrobots, particularly those located in inaccessible locations within the body, are generally restricted in terms of the available space available for power sources, thus power scavenging approaches are an attractive means to either solely power or, alternatively, extend the operational lifetime of these systems via a trickle-charge approach.

While trends show that the term energy harvesting has recently been adapted to refer to large-scale energy sources such as solar, wind, and wave/tidal energy, in many cases as commonly used in the literature the term suggests a kinetic mechanical-to-electrical energy conversion with small levels of energy harvested which, while not suitable for grid-level deployment, may find particular niche applications as a local, on-site power generation mechanism. When specifically considering vibration energy harvesting, the three most common approaches are to utilize piezoelectric, electromagnetic, or electrostatic transduction mechanisms to convert relative motion created by the external environment into electrical energy. A significant advantage of vibration energy harvesting is that relative motion and vibration are present in almost all environments, which is certainly the case in biological systems of interest in this book.

Although research in this area has exploded over the past several years and significant advances have been demonstrated in the research lab, a number of areas must still be addressed to facilitate the adaptation of this technology in practical applications. For example, resonance-based vibration energy harvesting approaches require techniques to ensure suitable bandwidth of the device (such that appreciable levels of energy are harvested over a range of input vibration source frequencies) [65, 66], while for optimal performance and reliability, the automated tuning of a vibrational energy harvesting device is required [67]. In addition, to realize vibrational energy harvesting devices as reliable power sources for MEMS-scale devices such as nano- and microrobotics requires an understanding of the lifetime stability and long-term performance of such devices. Here, a tightly coupled integration of the energy harvesting component with the larger overall system is clearly required to ensure optimal system performance. For example, one year of resonance at 200 Hz near the yield stress of the harvester will result in 10^9 high amplitude, high stress deformation cycles. This suggests the use of resonant frequency tunability to selectively tune to a specific frequency (not necessarily resonance) to balance the power output requirements with long-term life/performance of the harvesting device. Nonetheless, the convergence of advances in energy harvesting methodologies and low-power electronics with the exciting recent developments in the fabrication of multifunctional nanowires suggests that energy harvesting approaches show promise for the future powering of small-scale nano/microrobotics for biomedical applications.

Acknowledgments This work was supported in part by the National Science Foundation (Award No. CMMI-0846937). The authors would also like to thank Professor David Cappelleri at Stevens for his contribution to this work.

References

1. Steltz E, Seeman M, Avadhanula S, Fearing (2006) RS Power electronics design choice for piezoelectric microrobots. In: IEEE/RSJ international conference on intelligent robots and systems, 9–15 October 2006. pp 1322–1328. doi:10.1109/IROS.2006.281897
2. Albano F, Chung MD, Blaauw D, Sylvester DM, Wise KD, Sastry AM (2007) Design of an implantable power supply for an intraocular sensor, using POWER (power optimization for wireless energy requirements). J Power Sour 170(1):216–224. doi:10.1016/j.jpowsour.2007.04.007
3. Freitas RA Jr (2005) Current status of nanomedicine and medical nanorobotics. J Comput Theor Nanosci 2:1–25. doi:10.1166/jctn.2005.001
4. Köhler U, Kümpers J, Ullrich M (2002) High performance nickel-metal hydride and lithium-ion batteries. J Power Sour 105(2):139–144. doi:10.1016/s0378-7753(01)00932-6
5. Cook-Chennault KA, Thambi N, Sastry AM (2008) Powering MEMS portable devices—a review of non-regenerative and regenerative power supply systems with special emphasis on piezoelectric energy harvesting systems. Smart Mater Struct 17(4):043001. doi:10.1088/0964-1726/17/4/043001
6. Bandhauer TM, Garimella S, Fuller TF (2011) A critical review of thermal issues in lithium-ion batteries. J Electrochem Soc 158(3):R1–R25. doi:10.1149/1.3515880
7. Harb JN, LaFollette RM, Selfridge RH, Howell LL (2002) Microbatteries for self-sustained hybrid micropower supplies. J Power Sour 104(1):46–51. doi:10.1016/s0378-7753(01)00904-1
8. Nathan M, Golodnitsky D, Yufit V, Strauss E, Ripenbein T, Shechtman I, Menkin S, Peled E (2005) Three-dimensional thin-film Li-ion microbatteries for autonomous MEMS. J Microelectromech Syst 14(5):879–885. doi:10.1109/JMEMS.2005.851860
9. Peckerar M, Dilli Z, Dornajafi M, Goldsman N, Ngu Y, Proctor RB, Krupsaw BJ, Lowy DA (2011) A novel high energy density flexible galvanic cell. Energy Environ Sci 4(5):1807–1812. doi:10.1039/C1EE01075A
10. Dohle H, Mergel J, Stolten D (2002) Heat and power management of a direct-methanol-fuel-cell (DMFC) system. J Power Sour 111(2):268–282. doi:10.1016/s0378-7753(02)00339-7
11. Steinfeld G, Maru HC, Sanderson RA (1996) High efficiency carbonate fuel cell/turbine hybrid power cycle. In: Proceedings of the 31st intersociety energy conversion engineering conference (IECEC 96), 11–16 August 1996. vol 1122, pp 1123–1127. doi:10.1109/IECEC.1996.553865
12. Heinzel A, Hebling C, Müller M, Zedda M, Müller C (2002) Fuel cells for low power applications. J Power Sour 105(2):250–255. doi:10.1016/s0378-7753(01)00948-x

13. Stetten Fv, Kerzenmacher S, Lorenz A, Chokkalingam V, Miyakawa N, Zengerle R, Ducree J (2006) A one-compartment, direct glucose fuel cell for powering long-term medical implants. In: 19th IEEE international conference on micro electro mechanical systems, pp 934–937. doi:10.1109/memsys.2006.1627954
14. Davis F, Higson SPJ (2007) Biofuel cells—recent advances and applications. Biosens Bioelectron 22(7):1224–1235. doi:10.1016/j.bios.2006.04.029
15. Kerzenmacher S, Ducrée J, Zengerle R, Stetten Fv (2008) Energy harvesting by implantable abiotically catalyzed glucose fuel cells. J Power Sour 182(1):1–17. doi:10.1016/j.jpowsour.2008.03.031
16. Kerzenmacher S, Kräling U, Metz T, Zengerle R, von Stetten F (2011) A potentially implantable glucose fuel cell with Raney-platinum film electrodes for improved hydrolytic and oxidative stability. J Power Sour 196(3):1264–1272. doi:10.1016/j.jpowsour.2010.08.019
17. Sue C-Y, Tsai N-C (2012) Human powered MEMS-based energy harvest devices. Appl Energy 93:390–403. doi:10.1016/j.apenergy.2011.12.037
18. Bell LE (2008) Cooling, heating, generating power, and recovering waste heat with thermoelectric systems. Science 321(5895):1457–1461. doi:10.1126/science.1158899
19. Carmo JP, Goncalves LM, Correia JH (2010) Thermoelectric microconverter for energy harvesting systems. IEEE Trans Industr Electron 57(3):861–867. doi:10.1109/TIE.2009.2034686
20. Bottner H, Nurnus J, Gavrikov A, Kuhner G, Jagle M, Kunzel C, Eberhard D, Plescher G, Schubert A, Schlereth KH (2004) New thermoelectric components using microsystem technologies. J Microelectromech Syst 13(3):414–420. doi:10.1109/JMEMS.2004.828740
21. Massaro A, De Guido S, Ingrosso I, Cingolani R, De Vittorio M, Cori M, Bertacchini A, Larcher L, Passaseo A (2011) Freestanding piezoelectric rings for high efficiency energy harvesting at low frequency. Appl Phys Lett 98(5):053502–053503. doi:10.1063/1.3551725
22. Park JC, Park JY, Lee Y-P (2010) Modeling and characterization of piezoelectric d33-mode MEMS energy harvester. J Microelectromech Syst 19(5):1215–1222. doi:10.1109/JMEMS.2010.2067431
23. Lee BS, Lin SC, Wu WJ (2010) Fabrication and evaluation of a MEMS piezoelectric bimorph generator for vibration energy harvesting. J Mech 26(04):493–499. doi:10.1017/S172771910000469X
24. Elfrink R, Kamel TM, Goedbloed M, Matova S, Hohlfeld D, Andel YV, Schaijk RV (2009) Vibration energy harvesting with aluminum nitride-based piezoelectric devices. J Micromech Microeng 19(9): 094005. doi:10.1088/0960-1317/19/9/094005
25. Lee BS, Lin SC, Wu WJ, Wang XY, Chang PZ, Lee CK (2009) Piezoelectric MEMS generators fabricated with an aerosol deposition PZT thin film. J Micromech Microeng 19(6):065014. doi:10.1088/0960-1317/19/6/065014
26. Shen D, Park J-H, Noh JH, Choe S-Y, Kim S-H, Wikle Iii HC, Kim D-J (2009) Micromachined PZT cantilever based on SOI structure for low frequency vibration energy harvesting. Sens Actuators A Phys 154(1): 103–108. doi:10.1016/j.sna.2009.06.007
27. Shen D, Park J-H, Ajitsaria J, Choe S-Y, III HCW, Kim D-J (2008) The design, fabrication and evaluation of a MEMS PZT cantilever with an integrated Si proof mass for vibration energy harvesting. J Micromech Microeng 18(5): 055017. doi:10.1088/0960-1317/18/5/055017
28. Kim H, Bedekar V, Islam RA, Lee WH, Leo D, Priya S (2008) Laser-machined piezoelectric cantilevers for mechanical energy harvestingenergy harvesting. IEEE Trans Ultrason Ferroelectr Freq Control 55(9):1900. doi:10.1109/TUFFC.881
29. Marzencki M, Ammar Y, Basrour S (2007) Integrated power harvesting system including a MEMS generator and a power management circuit. Sens Actuators A 145–146:363–370. doi:10.1016/j.sna.2007.10.073
30. Beeby SP, Torah RN, Tudor MJ, Glynne-Jones P, O'Donnell T, Saha CR, Roy S (2007) A micro electromagnetic generator for vibration energy harvesting. J Micromech Microeng 17(7):1257. doi:10.1088/0960-1317/17/7/007
31. Fang H-B, Liu J-Q, Xu Z-Y, Dong L, Wang L, Chen D, Cai B-C, Liu Y (2006) Fabrication and performance of MEMS-based piezoelectric power generator for vibration energy harvesting. Microelectron J 37(11):1280–1284. doi:10.1016/j.mejo.2006.07.023

32. Jeon YB, Sood R, Jeong Jh, Kim SG (2005b) MEMS power generator with transverse mode thin film PZT. Sens Actuators A 122(1):16–22. doi:10.1016/j.sna.2004.12.032
33. Dutoit NE, Wardle BL, Kim S-G (2005) Design considerations for MEMS-scale piezoelectric mechanical vibration energy harvesters. Integr Ferroelectr Int J 71(1):121–160. doi:10.1080/10584580590964574
34. Williams CB, Yates RB (1996) Analysis of a micro-electric generator for microsystems. Sens Actuators A 52(1–3):8–11. doi:10.1016/0924-4247(96)80118-x
35. Glynne-Jones P, Tudor M, Beeby S, White N (2004) An electromagnetic, vibration-powered generator for intelligent sensor systems. Sens Actuators A 110(1–3):344–349. doi:10.1016/j.sna.2003.09.045
36. Saha CR, O'Donnell T, Loder H, Beeby S, Tudor J (2006) Optimization of an electromagnetic energy harvesting device. IEEE Trans Magn 42(10):3509–3511. doi:10.1109/TMAG.2006.879447
37. Serre C, Pérez-Rodríguez A, Fondevilla N, Morante JR, Montserrat J, Esteve J (2007) Vibrational energy scavenging with Si technology electromagnetic inertial microgenerators. Microsyst Technol 13(11–12):1655–1661. doi:10.1007/s00542-006-0338-1
38. Beeby SP, Tudor MJ, White NM (2006) Energy harvesting vibration sources for microsystems applications. Meas Sci Technol 17(12):R175. doi:10.1088/0957-0233/17/12/R01
39. Roundy S (2005) On the effectiveness of vibration-based energy harvesting. J Intell Mater Syst Struct 16(10):809–823. doi:10.1177/1045389x05054042
40. Meninger S, Mur-Miranda JO, Amirtharajah R, Chandrakasan AP, Lang JH (2001) Vibration-to-electric energy conversion. IEEE Trans Very Large Scale Integr (VLSI) Syst 9(1): 64–76. doi:10.1109/92.920820
41. Roundy S, Wright PK, Pister KSJ (2002) Micro-electrostatic vibration-to-electricity converters. In: ASME 2002 international mechanical engineering congress and exposition (IMECE2002). New Orleans, Louisiana, pp 487–496. doi:10.1115/IMECE2002-39309
42. Chiu Y, Kuo C-T, Chu Y-S (2007) MEMS design and fabrication of an electrostatic vibration-to-electricity energy converter. Microsyst Technol 13(11):1663–1669. doi:10.1007/s00542-006-0348-z
43. Bartsch U, Trautmann A, Ruther P, Gaspar J, Paul O (2007) Electrostatic transducers for micro energy harvesting based on SOI technology. In: Solid-state sensors, actuators and microsystems conference. pp 141–144. doi:10.1109/SENSOR.2007.4300091
44. Tashiro R, Kabei N, Katayama K, Ishizuka Y, Tsuboi F, Tsuchiya K (2000) Development of an electrostatic generator that harnesses the motion of a living body: use of a resonant phenomenon. JSME Int J Ser C 43(4):916–922
45. Han J, Von Jouanne A, Le T, Mayaram K, Fiez TS (2004) Novel power conditioning circuits for piezoelectric micro power generators. In: IEEE applied power electronics conference and exposition—APEC. pp 1541–1546. doi:10.1109/APEC.2004.1296069
46. Ivan IA, Rakotondrabe M, Agnus J, Bourquin R, Chaillet N, Lutz P, Poncot JC, Duffait R, Bauer O (2010) Comparative material study between PZT ceramic and newer crystalline PMN-PT and PZN-PT materials for composite bimorph actuators. Rev Adv Mat Sci 24(1–2):1–9
47. Sun C, Shi J, Bayerl DJ, Wang X (2011) PVDF microbelts for harvesting energy from respiration. Energy Environ Sci 4(11):4508–4512. doi:10.1039/C1EE02241E
48. Karami MA, Inman DJ (2012a) Powering pacemakers from heartbeat vibrations using linear and nonlinear energy harvesters. Appl Phys Lett 100(4):042901–042904. doi:10.1063/1.3679102
49. Karami MA, Inman DJ (2011) Analytical modeling and experimental verification of the vibrations of the zigzag microstructure for energy harvesting. J Vib Acoust 133(1):011002–011010. doi:10.1115/1.4002783
50. Karami MA, Inman DJ (2012b) Parametric study of zigzag microstructure for vibrational energy harvesting. J Microelectromech Syst 21(1):145–160. doi:10.1109/JMEMS.2011.2171321
51. Ramsay MJ, Clark WW (2001) Piezoelectric energy harvesting for bio-MEMS applications. In: McGowan A-MR (ed) Smart structures and materials 2001: industrial and commercial applications of smart structures technologies SPIE, pp 429–438. doi:10.1117/12.429684

52. Rocha JG, Goncalves LM, Rocha PF, Silva MP, Lanceros-Mendez S (2010) Energy harvesting from piezoelectric materials fully integrated in footwear. IEEE Trans Ind Electron 57(3). doi:10.1109/TIE.2009.2028360
53. Sodano HA, Inman DJ, Park G (2004) A review of power harvesting from vibration using piezoelectric materials. Shock Vib Dig 36(3):197–205. doi:10.1177/0583102404043275
54. Anton SR, Sodano HA (2007) A review of power harvesting using piezoelectric materials (2003–2006). Smart Mater Struct 16(3):R1. doi:10.1088/0964-1726/16/3/R01
55. Gao Y, Wang ZL (2009) Equilibrium potential of free charge carriers in a bent piezoelectric semiconductive nanowire. Nano Lett 9(3):1103–1110. doi:10.1021/nl803547f
56. Wang ZL, Song J (2006) Piezoelectric nanogenerators based on zinc oxide nanowire arrays. Science 312(5771):242–246. doi:10.1126/science.1124005
57. Wang ZL (2008) Towards self-powered nanosystems: from nanogenerators to nanopiezotronics. Adv Funct Mater 18(22):3553–3567. doi:10.1002/adfm.200800541
58. Yang R, Qin Y, Dai L, Wang ZL (2009) Power generation with laterally packaged piezoelectric fine wires. Nat Nanotechnol 4(1):34–39. doi:10.1038/nnano.2008.314
59. Zhu G, Yang R, Wang S, Wang ZL (2010a) Flexible high-output nanogenerator based on lateral ZnO nanowire array. Nano Lett 10(8):3151–3155. doi:10.1021/nl101973h
60. Yu A, Li H, Tang H, Liu T, Jiang P, Wang ZL (2011) Vertically integrated nanogenerator based on ZnO nanowire arrays. Phys Status Solidi Rapid Res Lett 5(4):162–164. doi:10.1002/pssr.201105120
61. Xu S, Shi Y, Kim S-G (2006) Fabrication and mechanical property of nano piezoelectric fibres. Nanotechnology 17(17):4497. doi:10.1088/0957-4484/17/17/036
62. Chen X, Xu S, Yao N, Shi Y (2010) 1.6 V nanogenerator for mechanical energy harvesting using PZT nanofibers. Nano Let 10(6):2133–2137. doi:10.1021/nl100812k
63. Chen X, Xu S, Yao N, Xu W, Shi Y (2009) Potential measurement from a single lead zirconate titanate nanofiber using a nanomanipulator. Appl Phys Lett 94(25):253113–253113. doi:10.1063/1.3157837
64. Li Y, Buddharaju K, Singh N, Lo GQ, Lee SJ (2011) Chip-level thermoelectric power generators based on high-density silicon nanowire array prepared with top-down CMOS technology. Electron Device Let 32(5):674–676. doi:10.1109/LED.2011.2114634
65. Zhu D, Tudor MJ, Beeby SP (2010b) Strategies for increasing the operating frequency range of vibration energy harvesters: a review. Meas Sci Technol 21(2):022001. doi:10.1088/0957-0233/21/2/022001
66. Challa VR, Prasad MG, Shi Y, Fisher FT (2008) A vibration energy harvesting device with bidirectional resonance frequency tunability. Smart Mater Struct 17(1):015035. doi:10.1088/0964-1726/17/01/015035
67. Challa VR, Prasad MG, Fisher FT (2011) Towards an autonomous self-tuning vibration energy harvesting device for wireless sensor network applications. Smart Mater Struct 20(2):025004. doi:10.1088/0964-1726/20/2/025004

Chapter 6
Capsule Robot in Gastro-Intestinal Tract: A Case Study for Robot Programming and Navigation

Shubo Zhang, Yi Guo and Arthur Ritter

Abstract We present a case study of a pill-sized capsule robot operating in the human's gastro-intestinal (GI) tract. A design example to conceptually build such a micro-robot is first presented, and a laboratory module is then developed to demonstrate robot navigation techniques. Medical considerations, such as size, speed, safety, and functionality of the robot are discussed, and robot building components are provided including sensors, actuators, processing, communicating, and power supply. The laboratory module is built on the 3D Webots simulation platform. Behavior-based robot navigation methods are introduced to program the robot to navigate in the human's body.

6.1 Introduction

It is estimated that 19 million people in the United States may suffer from diseases related to the small intestine, including obscure bleeding, irritable bowel syndrome, Crohn's disease, chronic diarrhea, and cancer. However, these pathologies are difficult to diagnose through traditional methods, such as push enteroscopy, wired endoscopes, and radiology [1]. Another disadvantage of traditional methods is that using an endoscope to inspect the inside of the human body is a rather uncomfortable procedure for a patient. Apart from that by using a conventional

S. Zhang · Y. Guo (✉)
Department of Electrical and Computer Engineering,
Stevens Institute of Technology, Hoboken, NJ 07030, USA
e-mail: yguo1@stevens.edu

A. Ritter
Department of Chemistry, Chemical Biology, and Biomedical Engineering,
Stevens Institute of Technology, Hoboken, NJ 07030, USA

Y. Guo (ed.), *Selected Topics in Micro/Nano-Robotics for Biomedical Applications*,
DOI: 10.1007/978-1-4419-8411-1_6, © Springer Science+Business Media New York 2013

endoscope there are potential side effects, such as perforation of organs, infection, and hemorrhage [2]. There are also various inspections of the digestive system such as the colon and stomach, and it is very difficult for the conventional endoscope to reach duodenum and small intestine.

6.1.1 Existing Capsule Endoscopy Technologies

Wireless capsule endoscopy has been commercially available since 2001. For a patient, such capsules offer a convenient examination with minimal preparation and immediate recovery [3]. The main vendors are Olympus Optical [4], Given Imaging [5], and the RF System Lab [6]. Given Imaging has developed two distinct capsules: PillCam ESO [7] for the esophagus and PillCam SB [8] for the small bowel. Both capsules measure 11 mm in diameter and 26 mm in length. RF data transmission at 433.10 MHz to an external antenna network enables the transfer of 2–8 frames per second. The device, powered by silver oxide batteries, can provide over 5 h of continuous video recording.

The RF System Lab has been developing the Norika3 system since 1998. Unlike PillCam, which uses complementary metal-oxide semiconductor (CMOS) image sensors, this device uses a charge-coupled device (CCD) image sensor. This results in superior image quality but with much greater power consumption due to the intense digital signal processing involved. To tackle the power requirement, the capsule transmits raw sensor data, while the processing, which consumes over 90 % of the power, occurs outside the body. The capsule, 9 mm in diameter and 23 mm long, is the smallest endoscopy capsule. It has four illumination LEDs with different light wavelengths. Three-dimensional coils within the capsule allow optimum power recovery from the inductive link. The system consists of the capsule; a vest with power transmission coils; a joystick-like device to control the capsule; and a PC system for signal processing, image display, and data storage. The second-generation capsule Sayaka [6], introduced in December 2005, operates on the same principle but has the lens on its lateral surface instead of its end. The imaging device is rotated within the capsule by steps of 7.5 degrees and provides approximately 30 frames per second, giving overall higher resolution.

The SmartPill Corporation has integrated temperature, pressure, and pH sensors into a single capsule, the Smart-Pill pH.p [9]. The company promotes the device as a complement to endoscopy with the potential to replace gastric emptying scintigraphy. The SmartPill GI Monitoring System includes the capsule, a wireless data receiver, a receiver docking station, and MotiliGI software. A powerful magnet activates an internal latching switch that provides a connection between the electronics and the battery. Once the capsule is activated, it begins working and transmits data to the mobile-phone-sized receiver (worn on the patient's belt). The receiver, in turn, transfers the data wirelessly to a PC in real time. The SmartPill, which has a 13-mm diameter and is 26 mm long, measures temperature to an accuracy of ±0.5 °C, pressure resolution to ±3.6 mm HG, and pH to ±0.28.

It uses the sensor data in addition to real and elapsed time measurements to provide gastric emptying time, combined small and large intestine transit time, contraction patterns, and a motility index. To optimize the device's performance, the pill samples at a high rate for the first 24 h and then at a decreased rate as the pill approaches the end of its journey. MotiliGI can plot the acquired data against time, providing invaluable information in the diagnosis of motility disorders such as gastroparesis (slowed passing of solids from the stomach). The pill underwent extensive clinical trials and consequently received FDA approval in July 2006.

Another interesting area of swallowable capsule technology is in vivo drug delivery or, conversely, sample extraction. Pharmaceutical Profiles is researching drug absorption using the patented Enterion capsule [10] developed by Phaeton Research. The capsule is 32 mm long and 11 mm in diameter. It can hold up to 1 ml of liquid or powder, which can expel at a target site in the body. The capsule contains a small amount of gamma-emitting tracer, allowing precise tracking in real time using an external gamma camera. When the capsule reaches the target area, an external electromagnetic field actuates the capsule's piston, ejecting the payload. The shell then passes harmlessly out of the body. Another piston-based capsule, proposed by the State University of New York, extracts substances from a target location in the gastro-intestinal (GI) tract in a similar fashion [11].

6.1.2 Robotic Capsule Technology

The commercial wireless capsule endoscopy we aforementioned has an advantage in reducing pain and discomfort of the patient due to its wireless nature. However, they must move passively from the mouth to anus by the peristaltic waves and gravity. If a capsule could be activated and controlled by a clinician to temporarily resist peristalsis and anchor itself in place in the gastrointestinal tract, additional tools for tissue biopsy, drug delivery, and cleaning or cauterizing angiectasias could be integrated into the device design to enhance its functionality. Due to the intestine's unstructured nature of the environment, which has loose, elastic, slippery walls [12], developing robotic capsules able to move inside the GI tract is more challenging. Nevertheless, recent research has been conducted on developing locomotion mechanisms into microcapsule robots. There are some capsule robots with anchoring mechanism that can resist the body's peristaltic forces to anchor itself to the intestinal lining at a desired location. Kim et al. [13] built and tested a capsule powered by eight individual shape memory alloy (SMA) actuators and embedded with microhooks to provide traction. Later, the same group developed a micromotor-actuated locomotion system for capsules inspired by canoe-paddling [14]. Quirini et al. [15] have proposed a pill-sized 12-legged endoscopic capsule for locomotion in the lower gastro-intestinal tract. Other locomotion methods that have been attempted for propulsion inside body cavities include a fin type electromagnetic actuator [16], a multi-joint endocavitary robot actuated by piezoelectric elements [17].

In this chapter, we present a case study on a pill-sized robot in the GI tract. The case study consists of a design example and a laboratory module. The design example proposes a conceptual design of a vitamin pill-sized robot vehicle that can operate within the human's GI tract. The laboratory module is based on the platform of the Webots simulator. The objective of the laboratory modules is to show how to program robots to navigate in an uncertain environment and how to control the robot. Two main robot navigation mechanisms will be demonstrated: the semi-autonomous and autonomous modes. In the semi-autonomous mode, humans can interfere with or control the robot through communication when needed; while in the autonomous mode, the robot is pre-programmed so that it achieves the task and adapts itself to the environment intelligently. In order to achieve autonomous navigation, behavior-based robotic navigation technique is first reviewed, and a few behaviors are created to program the robot navigation functionalities.

The rest of the chapter is organized as follows. In Sect. 6.2, we present a conceptual design of the pill-sized robot, and introduce its building blocks. In Sect. 6.3, we discuss the method to control the micro-robot navigating in the human's GI tract. Then in Sect. 6.4, a laboratory module is built to simulate different robot behaviors based on the Webots robot simulate platform. Finally, the chapter is concluded with brief remarks in Sect. 6.5.

6.2 Conceptual Design of Capsule Robot

6.2.1 Design Principles

Medical considerations provide design requirements for capsule robots, such as size, speed, safety, and functionality. The proposed capsule robot aims to effectively visualize the GI tract with navigation and tracking capability. The design principle of the capsule robot needs to consider both medical constraints and application scenarios.

Size: Since the size of the capsule is very important (very large pills make patients uncomfortable), the designed capsule must be sufficiently small to be swallowable. This limitation dramatically increases design difficulties. The trade-off between size and capabilities must be taken into consideration. The foremost challenging is miniaturization to obtain an ingestible device. In order to be swallowable, a capsule robot could fit within a cylindrical shape 9 mm in diameter and 23 mm long–the size of commercial pill-cameras such as the capsule Sayaka [6], is the smallest endoscopy capsule.

Speed: A standard colonoscopy is completed in approximately 20 min–1 h, so it is desirable for a locomoting robot to be able to move fast enough to travel through the colon in this time. While a fast response time would be preferable, it is impossible to immoderately increase speed because of the dissipation of power and patients' safety.

Safety: When operating within the human body, safety must be taken as the most important concern. The capsule's contact with the walls of the GI tract should cause no more damage than a standard colonoscope. The design of capsule must be considered with safety and biocompatibility.

Functionality: Some of the early concepts are now on the market. Families of sensing capsules provide temperature [9, 18], pressure [9], imaging [6, 7, 9, 10], pH data [9], drug delivery [10], tissue sampling [9], and polyps detection [4] to complement classic diagnostics, and one capsule delivers medication. There are also some potential applications, such as obscure bleeding, diagnostic pancreatic cancer, esophageal cancer, and gastric cancer. It is desirable to design a capsule robot that could combine the above functionalities and locomotion capabilities.

6.2.2 Application Scenarios

The proposed capsule is a self-contained micro-system that can perform sensing and actuating functions in the GI tract. Sensing function includes temperature, pressure, imaging, pH data, and tissue biopsy. Actuating function is that the capsule robot has the ability to move in 2-D, moving forward and turning, which enables it to implement tracking in the stomach. Since it is not economical to consider a medical doctor waiting all the time during the endoscopy process which can take around 5–20 h [19], the following scenario is proposed for the application of the robotic endoscopic capsule. At first, the capsule robot (with the tracking mode) is swallowed by the patient with the standard out of hospital process to get a map and the approximate positions of all GI tract regions with potential problems. In the tracking mode, the capsule robot could finish the GI tract faster than the standard commercial capsule camera. Next, the robotic capsule (with navigation mode) is swallowed by the patient. The capsule could be activated and controlled by a clinician to anchor itself in the desired place that needed to conduct a detailed temperature, imaging, pressure, pH data, and tissue biopsy. Locomotion capability of the capsule robot will allow the clinician to adjust the forward and turning position of the capsule. These operations could also be realized remotely by the doctor through the Internet.

6.2.3 Conceptual Design

Actuators: As outlined above, the actuator must be safe to use inside the human body. It would be preferable that the actuator has large strain capability, produces high output force, draws low power, and is small in size.

The selection of the actuators has been discussed in [20]. Piezoelectric materials were considered for their small size, high output force, quick speed, and low power consumption. However, the need for a high driving voltage raises questions

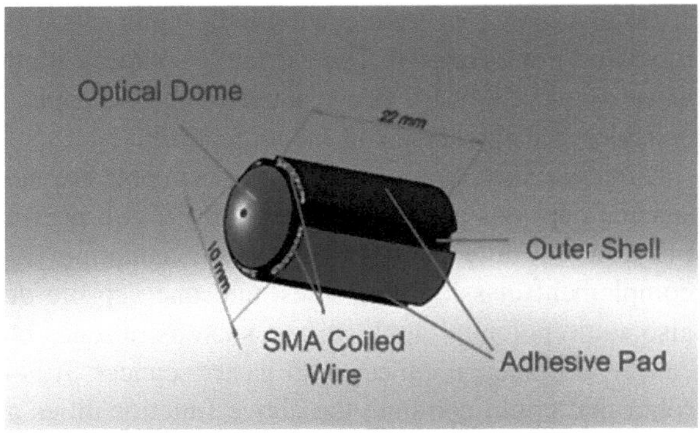

Fig. 6.1 The endoscope capsule robot

as to its biocompatibility. Combined with its small strain capability, this was determined to be a reasonable cause to look at other actuators.

Polymer actuators have become increasingly popular in robotics. The large strain capability and its biocompatibility are attractive. Unfortunately, these actuators are slow, relatively bulky, incapable of high output force, and consume large amounts of power.

The next actuator considered was SMA. This type of actuator had all the qualities necessary for this device with two exceptions. SMA is heat-activated and thus has very low efficiency and slow response time. In addition, this means that the device dissipates a lot of power. Despite this drawback, it was decided that SMA would be sufficient for the purposes of a capsule prototype. The issue of power consumption will be addressed as the project progresses further.

Our conceptual design of the capsule robot is inspired from the earthworm-like locomotive mechanisms proposed by Kim et al. [21]. In order to realize a 2-D locomotive mechanism, four spring-type SMA actuators are required to have long stroke and a strong enough force to overcome resistance force due to deformation of small intestine. The developed actuator is integrated with clampers mimicking claws of insects and an earthworm-like locomotive mechanism is proposed. The SMA actuators can be controlled to contract and stretch by passing current through the wire. When all four SMA are actuated in the same rhythm, the capsule robot moves forward or backward. Turning capability can be achieved by actuating the left and right SMAs in the opposite rhythm. Based on the design of actuators, the capsule robot has the ability to move in 2-D, moving forward and turning, which enables it to implement tracking and navigation in the GI tract.

The capsule robot measures 10 mm in diameter and 22 mm in length, see Fig. 6.1. The outer shell of the device is biocompatible material. The SMA coiled wire is attached to an adhesive pad. An optical dome is embedded in the front of the capsule. An inner shell contains five modules: vision module, sensors module, communication module, CPU module, and battery.

Vision Module: Unlike PillCam, which uses CMOS image sensors, this device uses a CCD image sensor. This results in superior image quality but with much

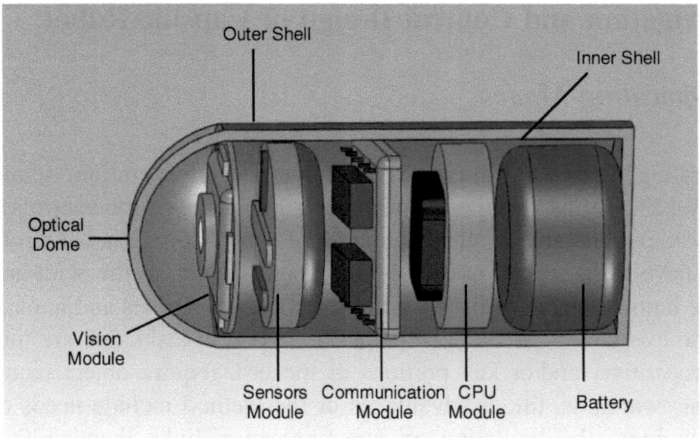

Fig. 6.2 Inside of the endoscope capsule robot

greater power consumption due to the intense digital signal processing involved. The CCD image sensor is compassed by four illumination light emitting diodes (LEDs) with different wavelengths.

Sensor Module: Sensors convert physical properties, such as light, pressure, or temperature into electrical signals. The capsule robot embeds sensors, including temperature, pressure, and pH data.

Communication Module: The communication module can then both transmit and receive the signal to communicate with the outside of the console. The RF antenna is utilized to receive external operation signal, such as activation, motion commands, and switch operation modes. Transmitter block sends the data, which is gathered from the sensors module, to the outside console.

CPU Module: The system's brain, the CPU, on one hand, digitizes the signals which are provided by the sensor and vision modules. On the other hand, the CPU performs additional processing of execution commands, which operates the SMA actuators in a controlled manner.

Power Supply: The capsule robot is powered by silver oxide batteries, which can provide over 5 h of continuous video recording. In battery-powered devices, the battery itself is likely the largest system component. Therefore, designers must minimize both supply voltage and current consumption while using high-efficiency topologies to achieve the required system performance.

As a conceptual design, one-third of the capsule will house the power supply and propulsion system, one-third will house the electronics including guidance, data transmission, and control, and one-third will house the hardware associated with sensing capabilities such as imaging, see Fig. 6.2.

6.3 Navigation and Control Design of Capsule Robot

6.3.1 Operating Modes

The operating modes of general robotics include teleoperation, semi- or fully autonomy [22]. In the teleoperation mode, a human operator controls the robot, who only views the environment through the robot's eyes, and the robot design does not have to figure out artificial intelligent. Instead, in the semi-autonomous mode, the human operator might control the robot sometimes and human does not have to do everything. While teleoperation is good at tasks that are unstructured and not repetitive, and/or key portions of the task require object recognition or situational awareness, the disadvantages of the method include needs of the display technology, the limitation on communication links (bandwidth and time delay), and the availability of trained personnel. On the contrast, routine or "safe" portions of the task are handled autonomously by the robot in the semi- or fully autonomous modes. Recent advances in the field of robotics have developed fully autonomous robots in various applications [23].

For our design example of pill-sized robot in the human's GI tract, the data rate can be dynamically adjusted by the following events: (1) change of sections in the GI tract (esophagus, stomach, small intestine, large intestine), (2) detection of tissue anomaly, (3) upon request by the physician. Taking into account the aforementioned events, we propose that the pill-sized robot operates in two working modes: semi-autonomous mode and autonomous mode.

Semi-autonomous Mode: In this mode, the robot can be activated and controlled by a clinician to anchor itself in the desired place to conduct a detailed sensing for temperature, imaging, pressure, pH data, or tissue biopsy. Locomotion capability of the robot will allow the clinician to adjust the forward and turning position of the capsule. These operations can also be realized remotely by the doctor through the Internet.

Autonomous Mode: In this mode, the pill-sized robot autonomously navigates in the GI tract. Most of the time, the robot utilizes the forward gait with a constant speed. When some predefined events occur, the robot will follow a predefined behavior. We introduce behavior-based robot programming in the next subsection.

6.3.2 Behavior-Based Robot Programming

Robot architecture determines how the robot takes sensor information, processes it, and takes actions accordingly. Robot architecture can be generally categorized as deliberative/hierarchical control, reactive behavior-based architecture, and hybrid systems which combines the former two methods [24]. The deliberative systems are hierarchical in structure with a clearly identifiable subdivision of functionality, where communication and control occur in a predictable and

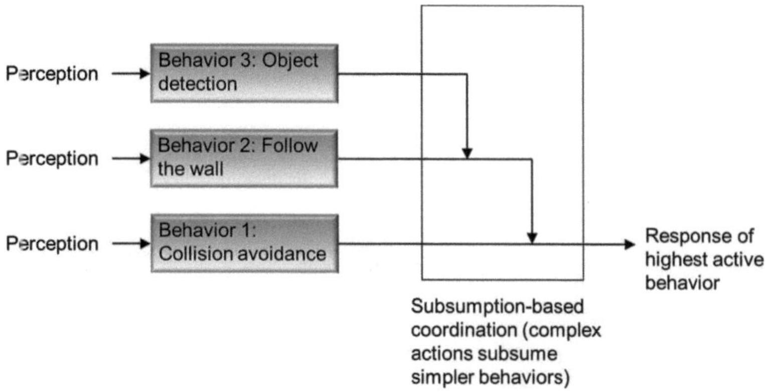

Fig. 6.3 Behavior-based robot programming for pill-sized robot navigation

predetermined manner followed up and down the hierarchy. It relies heavily on representations of world models. It is well suited for structured and highly predictable environments, but the drawbacks include slow actions due to model building and deliberate planning, the requirements of world modeling, and the limited communication pathways.

For unstructured and uncertain environments, reactive behavior-based methods are more effective. A behavior is a mapping of sensory inputs to a pattern of motor actions which then are used to achieve a task. Reactive control tightly couples perception and actions, typically in the context of motor behaviors, to generate timely robotic responses in dynamic environments [24]. Due to fast action and its robustness, it has been commonly used in robotics research and practices since its inception in the 1980s. The key aspect in the behavior-based control is how to coordinate behaviors. The popular subsumption architecture uses separated layers to represent individual goals which may happen concurrently and asynchronously. We adopt the behavior-based control in programming the micro-robot in the GI tract, and explain the method next.

For our design example of pill-sized robot in the human's GI tract, we program the following behaviors for the robots:

Behavior 1: Obstacle avoidance: It is the behavior to avoid potential collisions with obstacles in the environment. If the front sensor detects obstacles in front of the robot, the robot avoids the detected obstacle by circumnavigating it.

Behavior 2: Following the wall: It is the behavior that the micro-robot follows one side of the wall of the GI tract to navigate. The robot may operate under this behavior most of the time.

Behavior 3: Object detection: It is the behavior that the robot may slow down and take extensive pictures and sensing measurements when it detects inimical tissues. It may also issue an alert signal and send it to a human operator or a doctor for further checkup.

The above behaviors are organized in a subsumption-based coordination scheme, see Fig. 6.3, where the lower level behavior has a higher priority to be

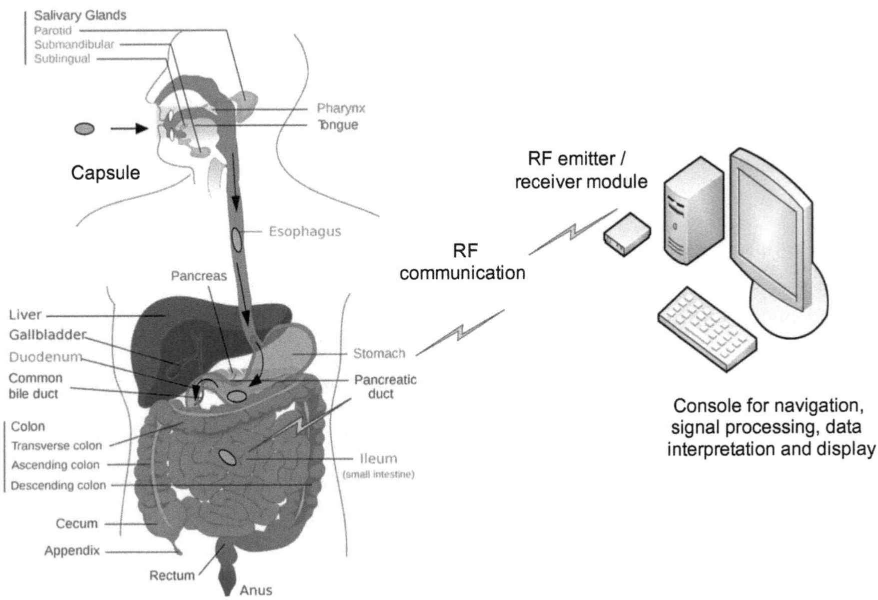

Fig. 6.4 An ingested capsule communicates wirelessly with an external control console

activated. That is, the robot may operate under "follow the wall" behavior when entering the GI tract; as soon as the sensor input indicates an "obstacles" on it way, the "obstacle avoidance" behavior is activated; and the "object detection" behavior is at the higher level, which is activated by the sensor input of "object". Note that the "obstacle" and "object" should be predefined for the robot, so that it can distinct between them real time through its sensor input.

6.4 Laboratory Module to Simulate Capsule Robot in GI Tract

The process of an ingested capsule communicating wirelessly with an external control console is shown in Fig. 6.4. The capsule robot is swallowed by the patient and traverses the esophagus, stomach, small intestines, and big intestine. During the robot navigation, the robot captures the information of the GI tract and transmits the data to the external console through the on-board communication module. Prior to the navigation, the external console may get a map of the GI tract and the approximate positions of the regions with potential problems in the tract. In the semi-autonomous mode, the clinician can drive the capsule robot to the desired places that are needed to acquire more detailed information.

To simulate the robot behaviors in the GI tract, we built a biomedical environment in the Webots simulator to imitate the GI tract. In the next, we first introduce the Webots robot simulator, and then present the robot simulation results.

6.4.1 Webots 3D Robotic Simulator

Webots is a software for fast prototyping and simulation of mobile robots. It has been developed since 1996 and was originally designed by Dr. Olivier Michel at EPFL, the Swiss Federal Institute of Technology in Lausanne, Switzerland, in the lab of Prof. Jean-Daniel Nicoud. Since 1998, Webots is a commercial product and is developed by Cyberbotics Ltd. This software has been used by over 750 universities and research centers worldwide. It offers a rapid prototyping environment, which allows the user to create 3D virtual worlds with physics properties, such as mass, joints, friction coefficients, etc. The user can add simple passive objects or active objects called mobile robots. These robots can have different locomotion schemes (wheeled robots, legged robots, or flying robots). Moreover, they may be equipped with a number of sensor and actuator devices, such as distance sensors, drive wheels, cameras, servos, touch sensors, emitters, receivers, etc. Finally, the user can program each robot individually to exhibit the desired behavior. Webots contains a large number of robot models and controller program examples to help users get started [25].

The graphic user interface (GUI) of Webots is composed of four principal windows: the 3D window that displays and allows to interact with the 3D simulation, the Scene tree which is a hierarchical representation of the current world, the Text editor allows to edit source code, and finally, the Console that displays both compilation and controller outputs, see Fig. 6.5.

A Webots simulation consists of the following components:

- A Webots world file that defines one or more 3D robot and their environment.
- Controller programs for the robots.
- An optional Supervisor.

A world, in Webots, is a 3D description of the properties of robots and of their environment. It contains a description of every object: its position, orientation, geometry, appearance (like color or brightness), physical properties, type of object, etc. Worlds are organized as hierarchical structures where objects can contain other objects (like in VRML97). For example, a robot can contain two wheels, a distance sensor and a servo which itself contains a camera, etc. A world file does not contain the controller code of the robots; it only specifies the name of the controller that is required for each robot. Worlds are saved in wbt files. The wbt files are stored in the worlds subdirectory of each Webots project.

A controller is a computer program that controls a robot specified in a world file. Controllers can be written in any of the programming languages supported by Webots: C, C++, Java, URBI, Python, or MATLAB. When a simulation starts, Webots launches the specified controllers, each as a separate process, and it associates the controller processes with the simulated robots. Note that several robots can use the same controller code; however, a distinct process will be launched for each robot. Some programming languages need to be compiled (C and C++), other languages need to be interpreted (URBI, Python and MATLAB), and some need to be

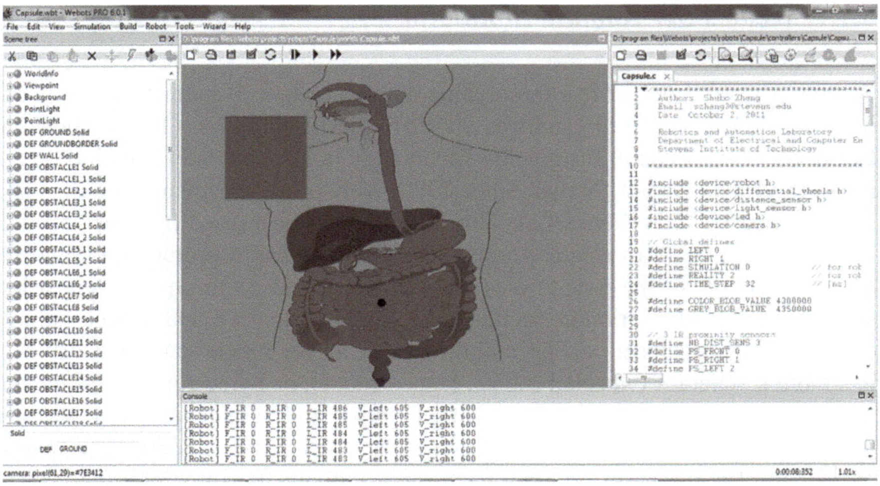

Fig. 6.5 Webots GUI composed of the scene tree (*left*), a display window (*top middle*), a text editor (*top right*), and the console window (*bottom*)

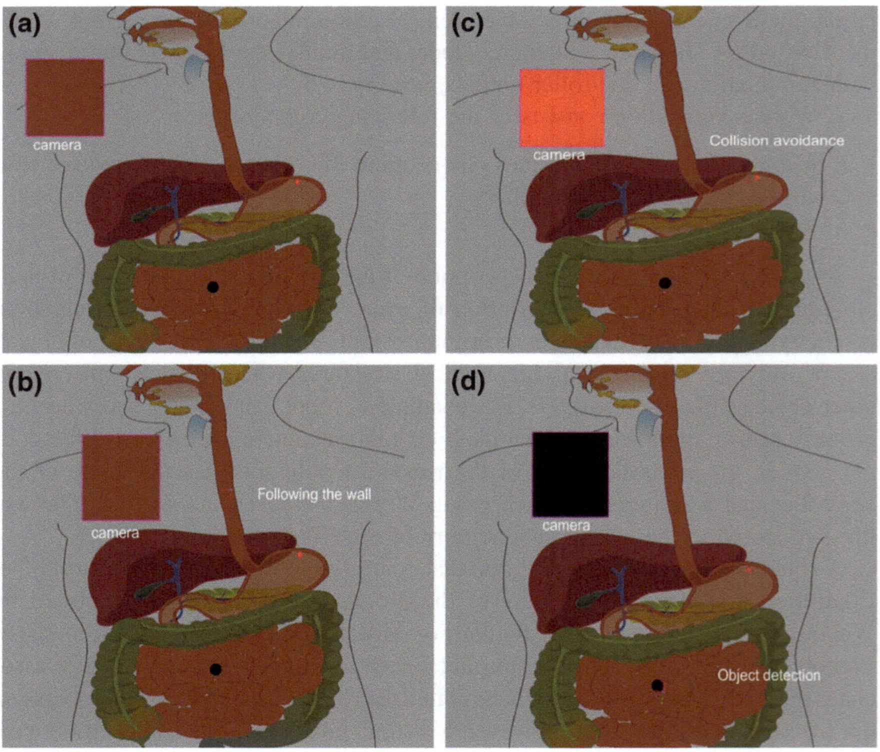

Fig. 6.6 The robot navigating in the GI tract: **a** The robot enters the GI tract; **b** Follow the wall behavior; **c** Collision avoidance behavior; **d** Object detection behavior

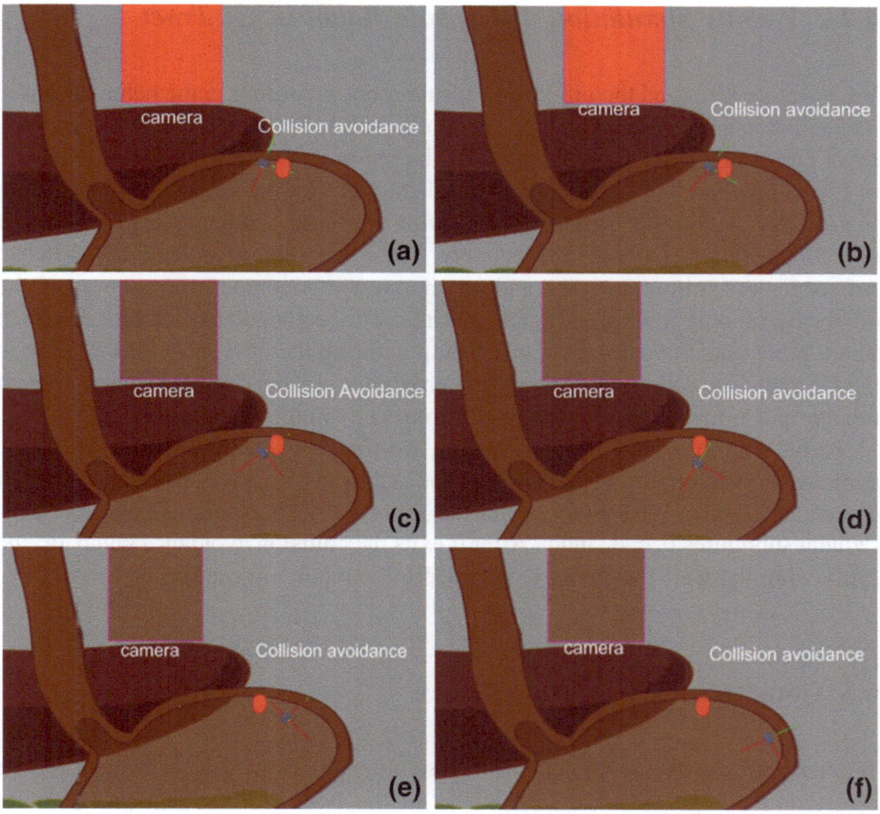

Fig. 6.7 Snapshots of the collision avoidance behavior: **a** The front camera detects the obstacle; **b** The robot starts to go around the obstacle; **c–e** The robot avoids the obstacle; **f** The robot resumes the follow the wall behavior

both compiled and interpreted (Java). For example, C and C++ controllers are compiled to platform-dependent binary executables (for example, .exe under Windows). URBI, Python, and MATLAB controllers are interpreted by the corresponding run-time systems (which must be installed). Java controllers need to be compiled to byte code (class files or jar) and then interpreted by a Java Virtual Machine. The source files and binary files of each controller are stored together in a controller directory. A controller directory is placed in the controllers subdirectory of each Webots project.

The Supervisor is a privileged type of Robot that can execute operations that can normally only be carried out by a human operator and not by a real robot. The Supervisor is normally associated with a controller program that can also be written in any of the above-mentioned programming languages. However, in contrast with a regular Robot controller, the Supervisor controller will have access to privileged operations. The privileged operation includes simulation control, for example, moving the robots to a random position, making a video capture of the simulation.

6.4.2 Webots Simulation of Capsule Robot in GI Tract

We have programmed the pill-sized capsule robot according to the behavior-based navigation and control design described in Sect. 6.3. We simulate a scenario of the robot passing through the GI tract under the pre-programmed behaviors, as shown in Fig. 6.6, where the robot enters the GI tract in Fig. 6.6a, and then navigate in Fig. 6.6b–d under the predefined following the wall, collision avoidance, and object-detection behaviors, respectively. Note that the robot can be switched between the semi-autonomous and autonomous modes upon the request of the doctor by sending a wireless signal anytime. The rectangular color block in each figure represents the real-time camera output during the navigation of the robot in the GI tract. The robot is equipped with proximity sensors (left and right) to detect relative positions with the walls, and a front camera to detect objects.

To have a close view of the behavior, we show the snapshots of the collision avoidance behavior in Fig. 6.7, where the onboard camera in front of the robot detects an obstacle in Fig. 6.7a, and the robot starts to go around it in the consequent subfigures b, c, d, and e. After it exits the collision avoidance behavior, the "following the wall" behavior is resumed to continue navigation.

6.5 Conclusion

Wireless capsule endoscopy represents a significant technical breakthrough for the investigation of the GI tract, especially in the light of disadvantages of other conventional techniques. Capsule endoscopy has the potential for use in a wide range of patients with a variety of illnesses. In this chapter, we present a conceptual design of a capsule robot navigating in the human's GI tract. Behavior-based programming was introduced to control the robot navigation in the human's body. A laboratory module was then developed to demonstrate the behaviors of the robot on the platform of the Webots simulator.

Acknowledgments Partial support for this work was provided by the National Science Foundation's Course, Curriculum, and Laboratory Improvement (CCLI) program under Award No. 0837584. Any opinions, findings, and conclusions or recommendations expressed in this material are those of the authors and do not necessarily reflect the views of the National Science Foundation.

References

1. Yu M (2002) M2A capsule endoscopy. A breakthrough diagnostic tool for small intestine imaging. Gastroenterol Nurs 25(1):24–27
2. Liedlgruber M, Uhl A (2009) Endoscopic image processing—an overview. In: Proceedings of 6th international symposium on image and signal processing and analysis, pp 707–712
3. Iddan G, Meron G, Glukhovsky A, Swain P (2000) Wireless capsule endoscopy. Nature 405:417

4. Olympus—Medical Systems & Endoscopy http://www.olympus-europa.com/endoscopy. Accessed 14 Feb 2012
5. Given Imaging Home—Gastrointestinal (GI) & Digestive Disease http://www.givenimaging.com. Accessed 14 Feb 2012
6. RF—The Next Generation Capsule Endoscope Sayaka http://www.rfsystemlab.com/en/sayaka/index.html. Accessed 14 Feb 2012
7. Device for In Vivo Imaging (2006) US Patent 7,009,634, to Given Imaging, Patent and Trademark Office
8. McCaffrey C, Chevalerias O, O'Mathuna C, Twomey K (2008) Swallowable-capsule technology. IEEE Pervasive Comput 7(1):23–29
9. SmartPill Corp http://www.smartpillcorp.com/index.cfm. Accessed 14 Feb 2012
10. Wilding I (2001) The enterion capsule: a novel technology for understanding the biopharmaceutical complexity of new molecular entities (NMEs). Drug Dev Deliv 1(1) (Posted on 3/28/2008)
11. Sampling Capsule and Process (1995) US Patent 5,395,366, to Gastrotarget Corp., Patent and Trademark Office
12. Phee L, Accoto D, Menciassi A, Stefanini C, Carrozza M, Dario P (2002) Analysis and development of locomotion devices for the gastrointestinal tract. IEEE Trans Biomed Eng 49(6):613–616
13. Kim B, Lee S, Park JH, Park JO (2005) Design and fabrication of a locomotive mechanism for capsule-type endoscopes using shape memory alloys (SMAs). IEEE/ASME Trans Mechatron 10(1):77–86
14. Park H, Park S, Yoon E, Kim B, Park J, Park S (2007) Paddling based microrobot for capsule endoscopes. In: Proceedings of the IEEE international conference on robot and automation, Rome, pp 3377–3382, April 10–14
15. Quirini M, Webster RJ III, Menciassi A, Dario P (2007) Design of a pill-sized 12-legged endoscopic capsule robot. In: Proceedings of the IEEE international conference on robot and automation, Rome, Italy, pp 1856–1862, April 10–14
16. Guo S, Sawamoto J, Pan Q (2005) A novel type of microrobot for biomedical application. In: IEEE international conference on intelligent robots and systems, pp 1047–1052
17. Yan G, Lu Q, Ding G, Yan D (2002) The prototype of a piezoelectric medical robot. In: Proceeding of IEEE international symposium on micromechatronics and human science, pp 73–77
18. Ingestible Thermometer Pill Helps Athletes Beat the Heat http://www.nasa.gov/vision/earth/technologies/thermometer_pill.html. Accessed 01 Aug 2010
19. Karagozler M, Cheung E, Kwon J, Sitti M (2006) Miniature endoscopic capsule robot using biomimetic micro-patterned adhesives. In: IEEE/RAS-EMBS international conference on biomedical robotics and biomechatronics, pp 105–111
20. Glass P, Sitti M, Appasamy R (2007) A new biomimetic adhesive for therapeutic capsule endoscope applications in the gastrointestinal tract. Gastrointest Endosc 65(5):AB 91
21. Kim B, Park S, Jee C, Yoon S (2005) An earthworm-like locomotive mechanism for capsule endoscopes. IEEE/RSJ international conference on intelligent robots and systems, pp 2997–3002
22. Murphy R (2000) Introduction to AI robotics. The MIT Press, Cambridge
23. Siciliano B, Khatib O (eds) (2008) Springer handbook of robotics. Springer-Verlag, Berlin
24. Arkin RC (1998) Behavior-based robotics. The MIT Press, Cambridge
25. Webots User Guide http://www.cyberbotics.com/guide. Accessed 14 Feb 2012

Chapter 7
Cooperative Control Design for Nanorobots in Drug Delivery

Shubo Zhang, Shuai Li and Yi Guo

Abstract In this chapter, we present cooperative control strategies for multi-robots to deliver drugs in tumor environments. We first discuss a nanorobot architecture, including chemical sensors, actuators, power supply, and data transmission, which is supported by the state of the art of nanotechnology. We then review tumor microenvironment modeling and pH measurement, where a tumor pH diffusion model is introduced and the pH value profile is established in the tumor environment. Based on the mathematical modeling, we propose a cooperative control strategy for pH sensitive nanorobots to deliver drugs in such environments. The control law is composed of gradient estimation and cooperative control, where the robots cooperatively estimate the gradient of the center of the group based on individual pH measurement, and then move towards the tumor center in a formation. We conduct rigorous convergence analysis and prove that the designed control steers the group of the robots reaching the cancer cells with the lowest pH value in the presence of estimation errors. Numerical simulations have shown effectiveness of the algorithm.

7.1 Introduction

A significant problem in cancer chemotherapy is the severe toxic side effects of anticancer drugs on healthy tissues. Invariably the side effects impose dose reduction, treatment delay, or discontinuance of therapy. It is challenging to limit

S. Zhang · S. Li · Y. Guo (✉)
Department of Electrical and Computer Engineering, Stevens Institute of Technology, Hoboken, NJ 07030, USA
e-mail: yguo1@stevens.edu

Y. Guo (ed.), *Selected Topics in Micro/Nano-Robotics for Biomedical Applications*,
DOI: 10.1007/978-1-4419-8411-1_7,

the adverse side effects of cancer chemotherapy on healthy organs and to strength drug efficiency to cancer. Researchers have developed a variety of strategies to get cancer drugs into the human body. There are two main approaches using nanoparticles: actively and passively targeted deliveries. Active targeting of a therapeutic agent is achieved by conjugating the therapeutic agent or carrier system to a tissue or cell-specific ligand [1]. Passive targeting is achieved by incorporating the therapeutic agent into a macromolecule or nanoparticle that passively reaches the target organ. However, neither the active nor the passive way is a perfect solution for drug delivery. They all face the same barrier: it is so difficult to navigate in the body's maze of vessels and tissues to seek out and destroy hidden cancers.

There are ongoing endeavors to build nanorobots for in vivo medical use. Nanorobots have been designed to travel through human blood vessels and microvasculatures to detecting and therapy diseases [2–7]. Many approaches to nanomedicine being pursued today show its successful development is almost inevitable [8]. The subsequent incorporation into valuable medical diagnostics or clinical therapeutics is highly possible and may occur very soon.

Some approaches to nanomedicine being pursued today are close to fruition. Cavalcanti et al. proposed several nanorobot architectures for medical nanorobotics [3–7, 9]. They presented an innovative hardware architecture for medical use of nanorobots, which were used as an advanced and precise tool for brain aneurysm diagnosis [3] and diabetes control [4]. Also, the proposed models offer details about how a nanorobot should help with the diagnosis of cerebral aneurysm and diabetes control based on clinical data and nanobioelectronics. In [5], the authors presented the robot architecture and the simulation of nanorobots with sensing capability for medical target identification. The proposed nanorobots can move with six-degree-of-freedom, and each nanorobot is guided by the sensor that has the ability to measure changes of chemical concentration without communications between robots.

Lewis and Bekey [10] have considered swarm intelligence and chemical signaling techniques for nanoscaled robots applied to the problem of tumor removal. Inspired by the concept from [10], Chandrasekaran and Hougen designed control strategies for bionanorobot swarm intelligence [11], where multi-directional sensing and motion strategy were adopted to achieve swarm intelligent behaviors to accomplish given tasks. In their setup, each nanorobot is capable of sensing the chemical signal in multiple directions, and making a decision on it own on the next move, and eventually reaching the target, which has the highest or the lowest concentration. Also, no communication occurs between robots.

In the above examples, nanorobots' group behaviors emerge from individual-level behaviors, where each agent measures the gradient of local environment and then follows the gradient. Simply speaking, in a 2D environment, the agent measures the concentration value of its front, left, and right directions through multi-directional sensors, and compares the signal strength, and then makes the decision for the next movement. The advantage of the method includes simplicity of the control law, but this control method may have slow response and low efficiency.

In this chapter, we propose a cooperative control method for nanorobots to deliver drugs. This is motivated by the success of plume tracing and source localization observed from chemotactic behaviors in the real world, e.g., bacteria are able to find food (e.g., glucose) by swimming towards the highest concentration of food molecules, or to flee from poisons (e.g., phenol) [12–14], which is a group-level behavior that individuals need local and their neighbors' information. In order to achieve similar performances to the observed chemotactic behaviors, we assume the nanorobots are able to communicate with their neighbors. Through cooperation, the nanorobots can drive the group to the tumor. Comparing to existing methods reviewed earlier, our proposed control strategy is a distributed control scheme where cooperating behaviors are achieved by robot-to-robot communications.

The rest of the chapter is organized as follows. In Sect. 7.2, we present a nanorobot architecture, including chemical sensor, actuator, power supply, and data transmission. In Sect. 7.3, we introduce pH selective therapy approaches and the techniques of pH imaging tumor. In Sect. 7.4, we discuss mathematical models of the pH distribution in the tumor microenvironment. The control strategy, which steers nanorobots tracing to the tumor target, is proposed in Sect. 7.5. In Sect. 7.5.3, we show simulation results of a group of nanorobots for tumor detection using the proposed cooperative control strategies. Finally, we conclude the chapter in Sect. 7.6.

7.2 Nanorobot Architecture

A medical nanorobot usually consists of a set of functionality blocks, including chemical sensor, actuator, communication, and power supply [2]. These modules make the nanorobot capable in a particular medical application. Each module should take into account the hardware size, as well as its applicability and biocompatibility for operation inside the human body.

Chemical Sensor. Nanosensors are used to convey information about nanoparticles in the microenvironment. By measuring changes in volume, concentration, displacement and velocity, gravitational, electrical and magnetic forces, pressure, or temperature of cells in a body, nanosensors may be able to distinguish between and recognize certain cells, most notably those of cancer, at the molecular level in order to deliver medicine or monitor development to specific places in the body [15].

Devices based on nanowires exhibit a number of key features, including label-fee, real-time electrical transduction, ultrahigh sensitivity and assembly of silicon circuits [16]. The ability of nanowire field effect devices to detect species in liquid solutions was demonstrated in 2001 for the case of hydrogen ion concentration or pH sensing [17]. Hahm and Lieber [18] investigated Si nanowire field-effect devices as sensors for the detection of single-stranded DNA. Nanowire devices can be the tool for drug discovery as well. In [19], the identification of molecular inhibitors to tyrosine kinases is discussed, which are proteins that mediate signal transduction in mammalian cells through phosphorylation of a tyrosine residue of a substrate protein using adenosine triphosphate (ATP). These examples manifest that nanowire devices have the potential to impact significantly on nanomedicine, such as disease diagnosis, genetic screening, and drug discovery, as well as to serve as powerful new tools for research in many areas of biology.

For medical nanorobots, the antibody CAB002167 is included for modeling the IC sensor which serves to identify higher concentrations of proteins that couple NOS isoforms to intracellular bloodstream signaling [20]. In [4], the human SGLT3 protein glucosensor [21] is embedded on nanorobot for glucose monitoring.

Actuator. The exceptional electromechanical characteristics of carbon nanotubes (CNTs) provide a promising avenue toward the development of actuators. A variety of properties of CNTs have been confirmed theoretically or experimentally. CNTs possess remarkable high electrical and thermal conductivies [22], high stiffness and strength [23], high aspect ratio (length/diameter) and low density [24]. These properties make them extraordinary as nanoelectronics wires and devices, such as nanowires [25], nanosensors [26], nanoactuators [27], nanotweezers [28], and memory system [29]. In [30], an array format based on CNTs and complementary metal oxide semiconductor (CMOS) techniques are proposed to achieve nanomanipulators as an embedded system for integrating nanodevices of molecular machines.

For medical nanorobots, it is indispensable that embedded actuators could be programmed to different behaviors, which enable nanorobots to operate inside the human body. Furthermore, precise trajectory motion and cooperative communication are essentially important for some biomedical problems, such as nanosurgery and intracellular drug delivery. In this Chapter, we discuss how to navigate nanorobots to locate tumor and then release drugs to destroy cancer cells. The nanorobots can only sense its local information, such as its current position and the pH value, but they do not know where the target is. Therefore, the most crucial problem is to find a path to reach the target by cooperation of the nanorobots.

Power Supply. Power is a major problem for nanorobots, which need energy to operate sensors, actuators and communications systems. A variety of power supply methods can be lumped into two categories, either from a source within the body or external to the body.

Molecule motor protein [31] and molecule motor DNA [32] nanodevices have been considered to perform transportation, chemical transformation, and enzymatic catalysis. The nanodevices might transduce chemical energy into mechanical motion and store it by using, for example, the biological energy currency of ATP, which can be envisaged.

The use of CMOS for active telemetry and power supply is probably the most effective and secure way to provide energy. Sauer et al. [33] proposed a radio frequency (RF) power harvesting telemetry chip, which is fabricated in 0.5 μm CMOS technology and supplies 1.7 mA at 3.3 V, over a distance up to 25 mm between coils, allowing the operation of several tasks with few or no significant losses during transmission.

Data Transmission. The use of radio frequency identification (RFID) based devices and sensors implanted inside the human body to transmit data about the health of patients offer many great advantages to study and monitor the human body. Mohseni et al. [34] proposed an integrated FM telemetry circuit for wireless in vivo recording of spontaneous neural activity in the very high frequency band of 94–98 MHz over a distance of 0.5 m. CMOS with sub-micrometer system-on-chip (SoC) design processes extremely low power consumption which could be used for nanorobot communicating through longer distances using acoustic sensors. For nanorobot communication, for navigational purposes, previous studies have shown that the use of nanoacoustics for nanorobot interactions can effectively achieve resolutions of 700 nm [35]. In [36], a sensing system for target localization and communication through active movement of an electric field emitter and sensor apparatus is designed. For the nanorobot active sonar communication, a chip is designed consuming less than 20 μW at 8 Hz repetition rate from a 3 V supply [37]. All above techniques provide practical ways for nanorobot active communication inside the human body.

For data transferring frequency, the IEEE standard of safety levels for human exposure to radio frequency electromagnetic fields is 3 KHz to 300 GHz [38]. Vaillancourt et al. [39] have shown that RF energy at 20 MHz penetrates the body with minimum energy loss and without any damage.

7.3 Tumor pH and Its Measurement

In this section, we review pH selective therapy approaches and techniques of pH imaging tumor.

Despite the general impression that the body has a uniform pH of 7.4, tumors tend to have lower pH values (as low as 5.7) [40]. In tumor cells, the increased glucose catabolism results in significant production of lactate and H^+. Although tumor cells have increased acid production, they maintain a normal or alkaline intracellular pH relative to normal cells. The major acid load is transported outside the cells, but the acid cannot be removed by the vasculature. The capacity of primary extracellular buffer is limited; as a result, the extracellular space becomes acidic. Changes in pH provide a selective tumor marker for diagnosis and therapeutic.

One approach to pH selective therapy is the use of agents which become active at low pH, as a result of either selective activation or pH dependent release. For instance, in tumor extracellular pH (pHe) targeting, Oh et al. [41] developed a pH responsive flower-like micelle to provide the mechanism for pH-triggered drug release from drug carriers. The micelle is stable at pH 7.0–7.4. As the pH decreases to slightly acidic pH (<pH 7.0), the micelles release the drug. Sethuraman et al. [42] designed a shielding/deshielding mechanism by positive

charges of TAT (Trans-Activator of Transcription), a cell penetrating peptide, on micelle surfaces controlled by the pH difference between 7.4 and 6.6. This shielding/deshielding mechanism suggests that an optimized pHe targeting system with an appropriate sulfonamide polymer is feasible. Wike-Hooley et al. [43] summarized the relevance between tumor pH and the treatment of malignant disease and found that low pH leads to decreased cell survival following treatment with hyperthermia, radiotherapy combined with hyperthermia, radio-sensitizers and various chemotherapeutic agents. Conversely, low pH affords some protection against radiation and some drugs.

A low extracellular pH is an important factor inducing more aggressive cancer phenotypes. Research into the causes and consequences of this acidic pH of tumors is highly dependent on accurate, precise, and reproducible measurements, and these have undergone tremendously changes in the last decade. Zhang et al. [44] gave a review which was in the in vivo measurement of tumor pH by pH sensitive PET radiotracers, MR spectroscopy, MRI, and optical imaging.

Pagel's group [45] has developed a paramagnetic Chemical Exchange Saturation Transfer agent, Yb (III)-1,4,7,10-tetraazacyclododecane-1,4,7,10-tetraacetic acid, o Aminoanilide (Yb-DO3A-oAA), with 2 pH-responsive CEST effects that have different MR frequencies and different dependencies on pH. The ratio of the 2 paramagnetic CEST effects can measure the entire physiologic range of pH from 6.1 to 8.0 with acceptable radiofrequency powers.

Bhujwalla et al. [46] have shown the pHe map of a 2 mm slice from an MDA-mb-435 breast cancer tumor in a severe combined immunodeficient (SCID) mouse. It generated from MRSI of the IEPA resonance, which was followed by a dynamic contrast enhancement series using Omniscan.

Garcia-Martin et al. [47] demonstrate that pH can be mapped in vivo in a rat brain glioma by H magnetic resonance spectroscopic imaging (MSI) of the pH buffer (+)2-imidazole-1-yl-3 ethoxycarbonylpropionic acid (IEPA). After infusion of IEPA, H spectra with (echo time) TE = 40 ms from within the glioma showed localization of the spectrum corresponds to the green point.

Techniques for measuring the pH of organelles, cytosol, and extracellular fluid in vivo have been consistently improved. Based on these techniques, designing pH sensitive nanorobots to detect tumor cells becomes feasible. In a tumor micro-environment, the pH distribution is measurable, so the nanorobots could track along gradient descending directions and locate tumor cells eventually. The diffusion model and tracking algorithms are discussed in the following sections.

7.4 The Tumor Microenvironment Modeling

As discussed in Sect. 7.3, a low extracellular pH is an important factor of physiological environment of a solid tumor. The excess H^+ ions diffuse along the concentration gradients from the tumor into adjacent normal tissues, resulting in a chronically acidic microenvironment for neighboring cells. Our cooperative control strategy will use this salient feature of the tumor microenvrionment.

In this section, we discuss mathematical models of the pH distribution in the tumor microenvironment. These models have been proposed by researchers who study acid-mediated tumor growth and invasion [48–52]. The study is based on a hypothesis that the transformed tumor metabolism with increased use of glycolysis and acid secretion alters the microenvironment by substantially reducing tumor extracellular pH. The ions produced by the tumor then diffuse along concentration gradients into the adjacent normal tissue. Acidification of the environment causes normal cell death. As a result, the tumor edge can be seen as forming a traveling wave progressing into the normal tissue, preceded by another traveling wave of increased microenvironmental acidity.

In the above literatures, the difference of the mathematical models is the additional factors, such as inhibition/stimulation tumor growth coefficient, production/induced death normal cells rate [50]. Gatenby et al. [48] proposed that tumor invasion of a normal tissue is primarily due to the low intracellular pH in the space surrounding a tumor. According to this model, the excess acid produced by cancer cells and the consequent decrease in intercellular pH leads to degradation of vital biochemical processes of the normal cells, and ultimately to their death. Their model assumes that tumors grow primarily by filling the hypo-cellular gap, created by the death of normal cells, through cell division. Tumor cells are assumed to be more resistant to low pH levels than normal cells, and thus can survive in the more acidic microenvironment created by cell necrosis and degradation. They extended the previous model through the inclusion of quiescent (non-proliferating) tumor cells since the previous models failed to capture key properties of the system, such as the existence of the benign steady state, or predicted incorrectly the size of the inter-tissue gap [49]. The improved model could fully identify the key parameters controlling different aspects of the behavior. Patel et al. [50] also extended the acid-invasion model developed in [48] to include both the competitive and cooperative interactions between the tumor and normal cells, by incorporating the influence of extracellular matrix and protease production at the tumor-stroma interface.

The model that we are going to use is based on the aforementioned hypothesis [51]. As assumed in [51], first, we assume that the tumor acts as an incompressible fluid. As such, local changes in the cell population, caused by the birth or death of the cells, give rise to the internal pressure gradients that induce cellular motion and the expansion or contraction of the tumor colony. Second, we model the tumor as a sphere and assume that spherical symmetry prevails at all times. A schematic cross-sectional view of a tumor and its surrounding normal tissue is given in Fig. 7.1. Let r_2 denote the tumor radius and r_1 the radius of the necrotic core. We assume that $r_1 < r < r_2$ is a viable region where the proliferating tumor cells exist in a spatially homogeneous state at their carrying capacity K_T. We further assume $r < r_1$ is a necrotic region, containing no viable cells, and that the necrotic debris continually disintegrates into simpler compounds that are freely permeable through cell membranes. The cell volume lost in this way is replaced by the cells pushed inward through adhesion or surface tension.

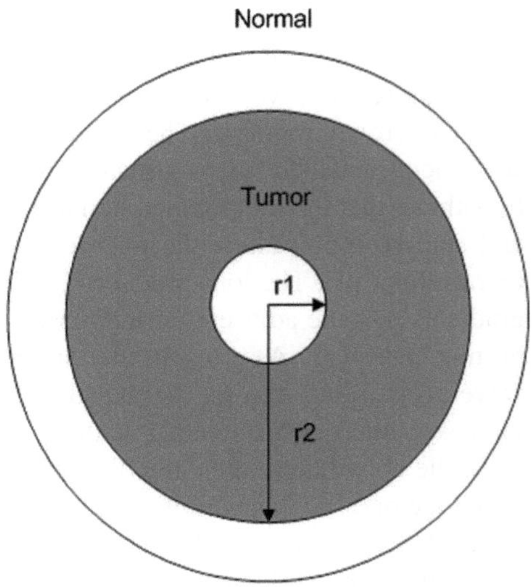

Fig. 7.1 The region of $r < r_1$ is a necrotic core. The region of $r_1 < r < r_2$ is the layer of proliferating tumor cells

Let H denote the extracellular concentration of excess hydrogen ions. Assuming that the evolution of H can be described by a reaction-diffusion equation [51]:

$$\frac{\partial H}{\partial t} = F_H + D_H \nabla^2 H \tag{7.1}$$

where D_H is the constant acid diffusion coefficient, and F_H represents the combined rate of acid production and removal from the system. Assume that H^* is the diffusive equilibrium and set $\partial H/\partial t = 0$ in the acid reaction-diffusion equation. Under these assumptions, and noting spherical symmetry, it becomes

$$0 = r_T T - r_V V H^* + \frac{D_H}{r^2}\frac{d}{dr}(r^2 \frac{dH^*}{dr}) \tag{7.2}$$

where r_T is the acid production rate of per unit volume of tumor, r_V is the acid removal rate proportional to the local acid concentration, T denotes the viable tumor cell density, and V the vascular density.

We define $V = 0$ for $r < r_2$ and $V = K_V$ elsewhere (avascular tumor case), i.e. there is no vasculature within the tumor and the vasculature exists homogeneously at its normal level outside the tumor. Taking tumor cell density T to be constant (K_T) within the viable region $r_1 < r < r_2$, and further taking $q = \sqrt{r_V K_V / D_H}$ and $H_0 = r_T K_T / r_V K_V$, we may normalize with $r = qR$ and $h = H/H_0$ to obtain:

$$rr^2 h'' + 2rh' = \begin{cases} 0 & 0 < r < r_1 \\ -r^2 & r_1 < r < r_2 \\ r^2 h & r_2 < r \end{cases} \tag{7.3}$$

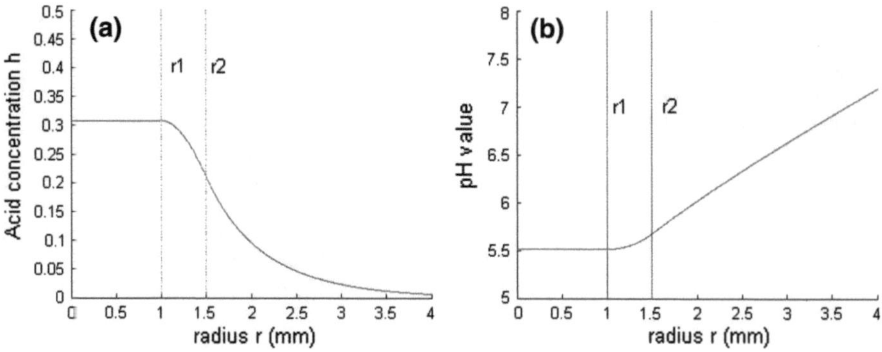

Fig. 7.2 **a** The acid concentration profile. **b** The pH value profile

where $h' = \frac{\partial h}{\partial r}$ and $h'' = \frac{\partial^2 h}{\partial r^2}$, with solution

$$h(r) = \begin{cases} k_1 & 0<r<r_1 \\ k_2 - k_3\frac{1}{r} - \frac{1}{6}r^2 & r_1<r<r_2 \\ k_4\frac{1}{r}e^{-r} & r_2<r \end{cases} \tag{7.4}$$

where $h(r)$ is the concentration of hydrogen ions as a function of r, the coefficients k_1, k_2, k_3, and k_4 are given below,

$$\begin{cases} k_1 = \frac{2r_1^3+3r_2^2+r_2^3}{6(r_2+1)} - \frac{r_1^2}{2} \\ k_2 = \frac{2r_1^3+3r_2^2+r_2^3}{6(r_2+1)} \\ k_3 = \frac{r_1^3}{3} \\ k_4 = \frac{e^{r_2}(r_2^3-r_1^3)}{3(r_2+1)} \end{cases} \tag{7.5}$$

Given experimentally determined parameter estimates of $q = 0.47\ \mathrm{mm}^{-1}$ and $H_0 = 1.0 \times 10^{-5}M \equiv$ pH 5.0 [48], this corresponds to a tumor of radius $r_2 = 1.5$ mm, with nectotic core radius $r_1 = 1$ mm. An example of this acid profile can be seen in Fig. 7.2a. The necrotic core region, $0<r<r_1$, is represented by a high acid concentration. From the viable region to the outside of tumor, $r_1<r$, acid concentration continues to decrease. Assume here that $\lim_{r\to\infty} h(r) = 0$, i.e., there is no excess acidity a long distance from the tumor.

The mathematical definition of pH is:

$$pH = \log 1/H \tag{7.6}$$

As we previously defined $h = H/H_0$, the pH value is shown in Fig. 7.2b. The region $0<r<r_1$ is the necrotic core of the tumor with the lowest pH value. From the viable region to the outside of the tumor, $r_1<r$, pH value continuous to increase. This corresponds to the change of the acid concentration h.

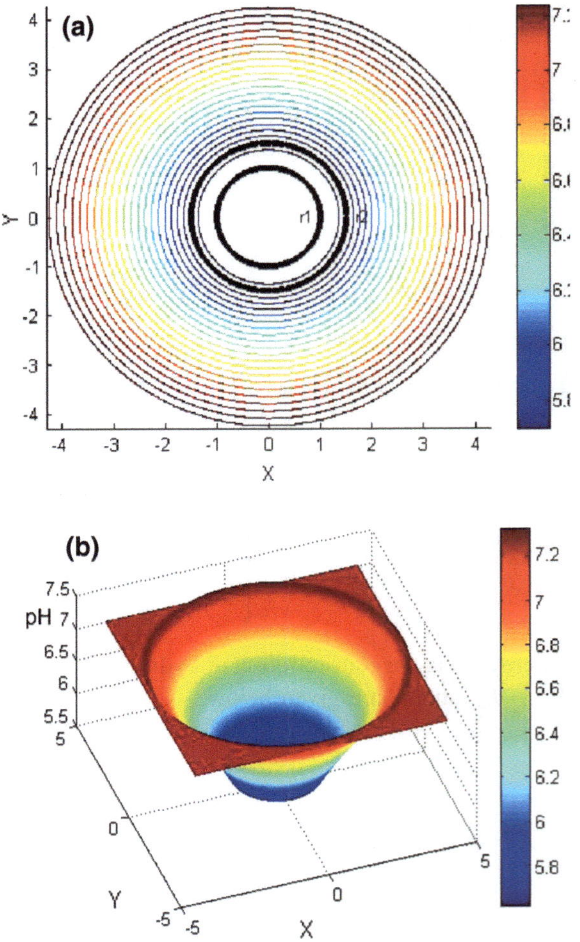

Fig. 7.3 **a** The tumor pH diffusion environment. **b** The variance of tumor pH value in a 3D plot. The *color bar* denotes the pH value defined in Eq. (7.6)

The tumor pH diffusion environment is depicted in Fig. 7.3a. It shows the cross section of the tumor and its surrounding tissues' pH, where the region of $r < r_1$, is the central necrotic core that has the lowest pH value. The pH value continues to increase outside the necrotic core. In order to illustrate clearly the variance of the tumor pH value, a 3D plot is given in Fig. 7.3b. In the necrotic core region, pH is around 5.6 (the lowest value).

7.5 Cooperative Control Strategy for Nanorobots

In this section, we discuss control strategies for a group of nanorobots to cooperatively deliver drugs in the tumor microenvironment described above.

We first have the following assumptions for the nanorobots:

- Each nanorobot has a pH sensor which can measure the pH value at the robot's current position.
- Each nanorobot can communicate with other nanorobots in the group.

Note that nanorobot communication techniques [35–37] have been reviewed in Sect. 7.2. Some mature techniques for pH measurement with accuracy, precision and high spatiotemporal resolution have been introduced in Sect. 7.3. Comparing to existing methods where each nanorobot measures the gradient of the local environment on its own, our proposed method uses a group of nanorobots to estimate local gradients of the pH value through neighboring robots' local samplings. Meanwhile, the nanorobots are expected to maintain a predefined formation which facilitates the estimation of the gradient. The estimated gradient of the pH value can drive the group of nanorobots to move to the tumor, which has a low pH value.

Another advantage of the multi-nanorobot strategy is that nanorobots may deliver sufficient therapy dose and facilitate to maintain a drug concentration. Furthermore, cancer cells are very stubborn and usually need more than one kind of drug for therapy. The multi-robot strategy satisfies this requirement, which can deliver multi-drugs to the destination.

7.5.1 Control Problem Formulation

We assume that the nanorobot's motion can be described by a double integrator:

$$\begin{aligned}\dot{\mathbf{x}}_i &= \mathbf{v}_i \\ \dot{\mathbf{v}}_i &= \mathbf{u}_i \quad \text{for } i = 1, 2, \ldots, n\end{aligned} \tag{7.7}$$

where $\mathbf{x}_i \in \mathbb{R}^k$, $\mathbf{v}_i \in \mathbb{R}^k$ and $\mathbf{u}_i \in \mathbb{R}^k$ are the position, the velocity and the control input (acceleration) of the ith nanorobot in a k dimensional workspace, respectively. In a compact matrix form, (7.7) writes,

$$\begin{aligned}\dot{\mathbf{x}} &= \mathbf{v} \\ \dot{\mathbf{v}} &= \mathbf{u}\end{aligned} \tag{7.8}$$

where $\mathbf{x} = [\mathbf{x}_1^T, \mathbf{x}_2^T, \ldots, \mathbf{x}_n^T]^T$, $\mathbf{v} = [\mathbf{v}_1^T, \mathbf{v}_2^T, \ldots, \mathbf{v}_n^T]^T$ and $\mathbf{u} = [\mathbf{u}_1^T, \mathbf{u}_2^T, \ldots, \mathbf{u}_n^T]^T$.

Our goal is to design a cooperative control strategy that drives the group of nanorobots to move to the tumor target while maintaining a desired formation. From this perspective, the nanorobot drug delivery problem can be regarded as cooperative source seeking problem as discussed in [53]. We formally define our control problem as follows.

Problem Definition: For a group of nanorobots R_1, R_2, ..., R_n locating at position $\mathbf{x}_1(t)$, $\mathbf{x}_2(t)$, ..., $\mathbf{x}_n(t)$, with the measurement of pH values, $pH(\mathbf{x}_1(t))$, $pH(\mathbf{x}_n(t))$, ..., $pH(\mathbf{x}_n(t))$, formed by a single tumor, design a distributed control law for each robot to drive all robots moving in a desired formation to the tumor region, that is, the region of $pH < pH^*$, where pH^* is the pH value at $r = r_2$, in the tumor microenvironment shown in Fig. 7.1.

Note that we define our control objective to be the group of robots reaching the tumor area, which is defined by certain pH value around the tumor area.

7.5.2 Cooperative Control Scheme

To develop control laws for the problem defined above, we follow the idea in our previous work [53] on source seeking, where the environment is modeled by a convex function. The pH value function (7.2), however, is not convex. Instead of using the pH value function directly, we define the following function, which is convex with respect to the spatial position $\mathbf{x}$:

$$p(r) = e^{pH} \tag{7.9}$$

Note that we use $p(r)$ as a "virtue" concentration function of the environment, and use it in the navigation.

Considering both (7.6) and (7.9), we get

$$pH(r) = \log 1/H = -\log h(r) - \log H_0 = -\log k_4 \frac{1}{r} e^{-r} - \log H_0 \qquad \text{for } r > r_2$$

where r is the distance to the tumor center. Therefore,

$$p(r) = e^{pH} = \frac{re^r}{k_4 H_0} \qquad \text{for } r > r_2$$

Note that

$$\frac{dp}{dr} = \frac{e^r(1+r)}{H_0 k_4} > 0 (r > r_2)$$

indicating that the function $p(r)$ is monotonically increasing. Also,

$$\frac{d^2p}{dr^2} = \frac{e^r(2+r)}{H_0 k_4} > 0 \qquad \text{for } r > r_2$$

indicating the function $p(r)$ is strictly convex for $r > r_2$. The function $p(\mathbf{x}) = p\big(r(\mathbf{x})\big)$ is a composition of the functions of $r(\mathbf{x})$ and $p(\mathbf{r})$. Since $r(\mathbf{x})$ is a convex function and $p(\mathbf{r})$ is both convex and monotonically increasing, $p(\mathbf{x})$ is a strictly convex function with respect to $\mathbf{x}$.

After defining the strict convex function $p(\mathbf{x})$, we can drive the robot following the gradient-descending direction to trace to the tumor. However, the gradient information is not available to each robot, and the robots need to estimate the gradient online during navigation. Therefore, to achieve the objective of driving nanorobots to the target tumor with a desired formation, the proposed control law consists of two parts: cooperative estimation of the gradient, and motion control with the estimated gradient. We will discuss these two parts in the next.

7.5.2.1 Cooperative Estimation of Gradients

Our goal is to estimate the gradient of the function $p(\mathbf{x})$ which is defined in (7.9) at the formation center of all nanorobots $\mathbf{x}_c(t) = \frac{1}{n}\sum_{i=1}^{n} \mathbf{x}_i(t)$, based on the sampling acquired by each nanorobot. To simplify the problem, we use a linear model to approximate $p(\mathbf{x})$ locally and adaptively update parameters of the linear model to gain a satisfactory estimation. In this way, the problem of estimating the gradient can be treated as an on-line parameter estimation problem by using a linear parameterization model [54]. The estimation model is as follows:

$$\hat{\mathbf{y}}(t) = [\mathbf{X}(t)\ \mathbf{1}]\theta(t) \tag{7.10}$$

with

$$\hat{\mathbf{y}}(t) = \begin{bmatrix} \hat{p}(\mathbf{x}_1(t)) \\ \hat{p}(\mathbf{x}_2(t)) \\ \dots \\ \hat{p}(\mathbf{x}_n(t)) \end{bmatrix}, \ \mathbf{X}(t) = \begin{bmatrix} \mathbf{x}_1^T(t) \\ \mathbf{x}_2^T(t) \\ \dots \\ \mathbf{x}_n^T(t) \end{bmatrix} \tag{7.11}$$

where $\mathbf{x}_1(t), \mathbf{x}_2(t), ..., \mathbf{x}_n(t)$ are all $k \times 1$ position vectors in the k-dimensional space, $\theta(t)$ is the estimation parameter, which is a $(k+1) \times 1$ vector, and $\mathbf{1}$ is a $n \times 1$ vector with all entries equal to 1. The estimation error can be described by the difference between the estimation $\hat{\mathbf{y}}(t)$ and the measurement $\mathbf{y}(t) = [p(\mathbf{x}_1(t)), p(\mathbf{x}_2(t)), ..., p(\mathbf{x}_n(t))]^T$. Least squares (LS) estimator can be employed to minimize the differences. We get the following parameter estimation by using a LS estimator:

$$\theta(t) = \begin{bmatrix} \mathbf{X}^T(t)\mathbf{X}(t) & \mathbf{X}^T(t)\mathbf{1} \\ \mathbf{1}^T\mathbf{X}(t) & \mathbf{1}^T\mathbf{1} \end{bmatrix}^{-1} \begin{bmatrix} \mathbf{X}^T(t) \\ \mathbf{1}^T \end{bmatrix} \mathbf{y}(t)$$

$$\hat{\mathbf{g}}_c(t) = \begin{bmatrix} \mathbf{I} & \mathbf{0} \end{bmatrix} \theta(t) \tag{7.12}$$

where $\hat{\mathbf{g}}_c(t)$ is the gradient estimation at the formation center $\mathbf{x}_c(t)$ at time t and $\mathbf{I}$ is a $k \times k$ identity matrix. This equation gives us an optimal estimation of $\hat{\mathbf{g}}_c(t)$ in terms of least squares.

7.5.2.2 Cooperative Control Law

The multi-robot cooperative control is composed of two parts: inter-robot artificial potential and source attraction potential. The inter-robot artificial potential is defined between neighboring nanorobots and accounts for the formation maintenance, which are repulsive when the robot is closer to its neighbors than the distance in the desired formation, while attractive when the robot is farther away from its neighbors than the desired distance. The source attraction potential is the potential attracting the robot to the source (i.e., tumor). Following the gradient descending direction, the nanorobot is able to reach the targeted tumor.

Using the estimated gradient (7.12), we present the following control law for each nanorobot:

$$\begin{aligned}\mathbf{u}_i = &- \sum_{j\in\mathbb{N}(i)} \omega_{1ij}(\mathbf{x}_i - \mathbf{x}_j - \mathbf{x}_{vi} + \mathbf{x}_{vj}) + c_0\hat{\mathbf{g}}_c \\ &- \sum_{j\in\mathbb{N}(i)} \omega_{2ij}(\mathbf{v}_i - \mathbf{v}_j) - \frac{c_1}{n}\sum_{i=1}^{n}\mathbf{v}_i - c_2\ \mathrm{sgn}(\sum_{i=1}^{n}\mathbf{v}_i)\end{aligned} \tag{7.13}$$

where $\mathbf{x}_{vi}$ and $\mathbf{x}_{vj}$ are desired relative positions of the ith and the jth robots in the desired formation, respectively; $\omega_{1ij} > 0$, $\omega_{2ij} > 0$ for $j \in \mathbb{N}(i)$, and $\omega_{1ij} = 0$, $\omega_{2ij} = 0$ for $j \notin \mathbb{N}(i)$, which define the weights of the links between the ith and the jth robots.

The control law (7.13) can be written in a compact form for all nanorobots,

$$\begin{aligned}\bar{\mathbf{u}} = &- (\mathbf{L}_1 \otimes \mathbf{I})(\bar{\mathbf{x}} - \bar{\mathbf{x}}_v) - (\mathbf{L}_2 \otimes \mathbf{I})\dot{\bar{\mathbf{x}}} + c_0\mathbf{1} \otimes \hat{\mathbf{g}}_c \\ &- \frac{c_1}{n}\mathbf{1} \otimes \left((\mathbf{1}^T \otimes \mathbf{I})\dot{\bar{\mathbf{x}}}\right) - c_2\mathbf{1} \otimes\ \mathrm{sgn}\left((\mathbf{1}^T \otimes \mathbf{I})\dot{\bar{\mathbf{x}}}\right)\end{aligned} \tag{7.14}$$

where $\bar{\mathbf{x}} = [\mathbf{x}_1^T, \mathbf{x}_2^T, ..., \mathbf{x}_n^T]^T$ is the position vector, $\bar{\mathbf{x}}_v = [\mathbf{x}_{v1}^T, \mathbf{x}_{v2}^T, ..., \mathbf{x}_{vn}^T]^T$ is the desired formation vector, both $\mathbf{L}_1$ and $\mathbf{L}_2$ are symmetric Laplacian matrices [1] on the undirected graph constructed by the group of robots, $\mathbf{I}$ is an $k \times k$ identity matrix, n is the number of robots, k is the dimension of the space, $\mathbf{1}$ is a $n \times 1$ vector with all entries equal to 1, c_0 is a negative constant, c_1 and c_2 are positive constants, $\mathrm{sgn}(\cdot)$ is the sign function, which equals to 1, -1 and 0 for a positive input, negative input, and the input of 0, respectively, and $\otimes$ is the Kronecker product.

The control law (7.14) consists of the following terms: formation term $-(\mathbf{L}_2 \otimes \mathbf{I})\dot{\bar{\mathbf{x}}}$, relative velocity damping term $(\mathbf{L}_1 \otimes \mathbf{I})(\bar{\mathbf{x}} - \bar{\mathbf{x}}_v)$, gradient descent term $c_0\mathbf{1} \otimes \hat{\mathbf{g}}_c$, linear absolute velocity damping term $-\frac{c_1}{n}\mathbf{1} \otimes \left((\mathbf{1}^T \otimes \mathbf{I})\dot{\bar{\mathbf{x}}}\right.$, and nonlinear absolute velocity damping term $-c_2\mathbf{1} \otimes\ \mathrm{sgn}\left((\mathbf{1}^T \otimes \mathbf{I})\dot{\bar{\mathbf{x}}}\right.$. The formation term accounts for the formation feedback control, and the relative velocity damping term attenuates the relative movement between nanorobots to achieve consensus of relative position. The gradient descent term guides nanorobots to

[1] The ith row and jth column of the Laplacian matrix L_{1ij} is defined as $L_{1ij} = -\omega_{1ij}$ with $\omega_{1ij} = 0$ for $j \notin \mathbb{N}(i)$, $\omega_{1ij} > 0$ for $j \neq i, j \in \mathbb{N}(i)$, and $L_{1ij} = \sum_{l\in\mathbb{N}(i)} \omega_{1il}$ for $j = i$. The ith row and jth column of the Laplacian matrix L_{2ij} is defined as $L_{2ij} = -\omega_{2ij}$ with $\omega_{2ij} = 0$ for $j \notin \mathbb{N}(i)$, $\omega_{2ij} > 0$ for $j \neq i, j \in \mathbb{N}(i)$, and $L_{2ij} = \sum_{l\in\mathbb{N}(i)} \omega_{2il}$ for $j = i$.

move towards the tumor. The rest two terms can decelerate nanorobots when they approach the tumor. Note that, the nonlinear absolute velocity damping term is robust against the noise and works well in the case with estimation error.

We present the control algorithm in Algorithm 1, where the ith robot collects its measurement of pH value pH_i, position $\mathbf{x}_i$ and velocity $\mathbf{v}_i$ (Line 2), and receives pH value, position and velocity of all the other robots by communication (Line 3). After this, the value of $p(\cdot)$ and the position matrix $\mathbf{X}$ are calculated according to Eq. (7.9) and (7.11) (Lines 4 and 5). Then, the gradient estimation of $\hat{\mathbf{g}}_c(t)$ is calculated according to Eq. (7.12) (Line 6). Subsequently, Eq. (7.13) is used to calculate the control input $\mathbf{u}_i$ for robot i. Lines 2, 3, 4, 5, 6, 7 are repeated in sequence until one or more nanorobots has a measurement of the pH value less than the threshold pH^* (pH^* is defined as the pH value at $r = r_2$), which indicates that the robot reaches the tumor area.

Note that the above algorithm represents an all-to-all communication scenario, where each robot collects all other robots' sensing information. The proposed control scheme can be extended to the case with limited communication. Interested readers are referred to [53].

Algorithm 1 Control algorithm for the ith nanorobot

Require:
pH value measurement pH, position $\mathbf{x}$, velocity $\mathbf{v}$ and desired formation $\mathbf{x}_v$ of all robots are available to the ith robot.
A pre-defined desired formation.

Ensure:
To drive the formation center to the tumor region with the pre-defined formation.

1: **repeat**
2: $pH_i, \mathbf{x}_i, \mathbf{v}_i \Leftarrow$ Sensor readings.
3: $pH_i, \mathbf{x}_i, \mathbf{v}_i$ for all i $\Leftarrow$ Communication with all the other robots.
4: p_i for all $i \Leftarrow$ Equation (7.9).
5: Position matrix $\mathbf{X} \Leftarrow$ Equation (7.11).
6: Gradient estimation $\hat{\mathbf{g}}_c(t) \Leftarrow$ Equation (7.12).
7: Control input $\mathbf{u}_i \Leftarrow$ Equation (7.14).
8: **until** (for some i $pH_i < pH^*$)

7.5.2.3 Convergence Analysis

In this section, we conduct convergence analysis of the proposed strategy under the assumption that the gradient estimation error is bounded by e_0, i.e., $\|\hat{\mathbf{g}}_c(t) - \mathbf{g}_c(t)\| \leq e_0$. The key part in the analysis is to prove the *attraction* of the robot group to a neighborhood of the tumor center, after which the exit criteria of the Algorithm stops the program when the robots are within the attraction region.

As the control objective is to drive nanorobots to the tumor region with $r < r_2$, as shown in Fig. 7.1, we need to design control laws so that the robots are attracted towards the tumor area when they are outside of it. To facilitate theoretical analysis, we assume a virtual center, $\mathbf{x}^*$, with a uniform concentration distribution the same as the one outside of the region $r<r_2$, and prove that the robot group is attracted to $\mathbf{x}^*$. There are three steps for the analysis of this part: Step 1: We first prove that the relative positions of nanorobots reach the desired formation, and the position of their formation center moves to an invariant set. Step 2: Based on the result obtained in Step 1, we then prove that the ultimate position of the formation center is within a bounded area from the center of the tumor. Step 3: Finally, we prove that the bounded region is within the tumor area.

Step 1: Substituting the control input (7.14) into the robot dynamics (7.7) yields

$$\begin{aligned}\ddot{\bar{\mathbf{x}}} &= -(\mathbf{L}_1 \otimes \mathbf{I})(\bar{\mathbf{x}} - \bar{\mathbf{x}}_v) - (\mathbf{L}_2 \otimes \mathbf{I})\dot{\bar{\mathbf{x}}} + c_0 \mathbf{1} \otimes \hat{\mathbf{g}}_c \\ &\quad - \frac{c_1}{n} \mathbf{1} \otimes \left((\mathbf{1}^T \otimes \mathbf{I})\dot{\bar{\mathbf{x}}}\right) - c_2 \mathbf{1} \otimes \ \mathrm{sgn}\left((\mathbf{1}^T \otimes \mathbf{I})\dot{\bar{\mathbf{x}}}\right)\end{aligned} \tag{7.15}$$

This equation represents the close-loop dynamics of the nanorobots.

As to the formation center $\mathbf{x}_c = \frac{1}{n}\sum_{i=1}^{n} \mathbf{x}_i$, its dynamics can be obtained as follows

$$\ddot{\mathbf{x}}_c = c_0 \hat{\mathbf{g}}_c - c_1 \dot{\mathbf{x}}_c - c_2 \ \mathrm{sgn}(\dot{\mathbf{x}}_c) \tag{7.16}$$

We choose the following Lyapunov function

$$\begin{aligned}V &= (\bar{\mathbf{x}} - \bar{\mathbf{x}}_v - \mathbf{1} \otimes \mathbf{x}_c)^T (\mathbf{L}_1 \otimes \mathbf{I})(\bar{\mathbf{x}} - \bar{\mathbf{x}}_v - \mathbf{1} \otimes \mathbf{x}_c) \\ &\quad + (\dot{\bar{\mathbf{x}}} - \mathbf{1} \otimes \dot{\mathbf{x}}_c)^T (\dot{\bar{\mathbf{x}}} - \mathbf{1} \otimes \dot{\mathbf{x}}_c) - p(\mathbf{x}_s) + p(\mathbf{x}_c) - \frac{1}{2c_0} \dot{\mathbf{x}}_c^T \dot{\mathbf{x}}_c\end{aligned} \tag{7.17}$$

where $\mathbf{x}_s = \ \mathrm{argmin}\{p(\mathbf{x})\}$. Note that the last term is positive as c_0 is negative. This Lyapunov function is analogous to an energy function, which provides a metric to measure the energy conserved by the system. The system is convergent if the Lyapunov function keeps decreasing in time, which is analogous to a system with energy dissipation. Now, we study the time derivative of the Lypunov function. By calculating the time derivative of V along the trajectory of (7.15), we can finally get

$$\dot{V} = -2\dot{\bar{\mathbf{x}}}^T (\mathbf{L}_2 \otimes \mathbf{I})\dot{\bar{\mathbf{x}}} + (\hat{\mathbf{g}}_c - \mathbf{g}_c)^T \dot{\mathbf{x}}_c + \frac{c_1}{c_0} \dot{\mathbf{x}}_c^T \dot{\mathbf{x}}_c + \frac{c_2}{c_0} \|\dot{\mathbf{x}}_c\|_1 \tag{7.18}$$

where $\|\cdot\|_1$ denotes the 1-norm of a vector. Using the properties of vector norms, we further get

$$\begin{aligned}\dot{V} \leq & -2\dot{\bar{\mathbf{x}}}^T (\mathbf{L}_2 \otimes \mathbf{I})\dot{\bar{\mathbf{x}}} + \frac{c_1}{c_0} \dot{\mathbf{x}}_c^T \dot{\mathbf{x}}_c + (\frac{c_2}{c_0} + \sqrt{k} e_0) \|\dot{\mathbf{x}}_c\|_1 \\ \leq & 0\end{aligned} \tag{7.19}$$

where $\frac{c_2}{c_0} + \sqrt{k} e_0 < 0$. We have,

$$\dot{V} = 0 \Rightarrow \dot{\bar{\mathbf{x}}} = \mathbf{1} \otimes \alpha_1 \ \text{ and } \dot{\mathbf{x}}_c = \mathbf{0} \tag{7.20}$$

where α_1 is a vector with k rows. Further, we have

$$\dot{V} = 0 \Rightarrow \dot{\mathbf{x}}_c = \mathbf{0} \tag{7.21}$$

$$\dot{V} = 0 \Rightarrow (\mathbf{L}_1 \otimes \mathbf{I})(\bar{\mathbf{x}} - \bar{\mathbf{x}}_v) = 0 \tag{7.22}$$

According to the invariance-like theorem [55], we conclude that (7.21) implies the velocity of the formation center reaches zero ultimately, while (7.22) implies the desired formation $\bar{\mathbf{x}}_v$ is reached ultimately. With the dynamics of the formation center, we further conclude that $\hat{\mathbf{g}}_c \to 0$.

Step 2: The Taylor expansion of the convex function $p(\mathbf{x})$ at position $\mathbf{x}^*$, where $\mathbf{x}^*$ represents the ultimate position of the formation center with $\hat{\mathbf{g}}_c(\mathbf{x}^*) = 0$, is

$$p(\mathbf{x}) = p(\mathbf{x}^*) + \nabla^T p(\mathbf{x}^*)(\mathbf{x} - \mathbf{x}^*) + \frac{1}{2}(\mathbf{x} - \mathbf{x}^*)^T H(\mathbf{x}_1)(\mathbf{x} - \mathbf{x}^*) \tag{7.23}$$

where $\mathbf{x}_1$ is between $\mathbf{x}$, and $\mathbf{x}^*$ and $H(\mathbf{x}_1)$ is the Hessian matrix at $\mathbf{x}_1$. For $\mathbf{x} = \mathbf{x}_s = \operatorname{argmin}\{p(\mathbf{x})\}$, this equation yields

$$p(\mathbf{x}_s) = p(\mathbf{x}^*) + \nabla^T p(\mathbf{x}^*)(\mathbf{x}_s - \mathbf{x}^*) + \frac{1}{2}(\mathbf{x}_s - \mathbf{x}^*)^T H(\mathbf{x}_1)(\mathbf{x}_s - \mathbf{x}^*) \tag{7.24}$$

Thus, we get

$$\begin{aligned} p(\mathbf{x}_s) - p(\mathbf{x}^*) &> -\|\nabla p(\mathbf{x}^*)\|\|\mathbf{x}_s - \mathbf{x}^*\| + \frac{\xi_1}{2}\|\mathbf{x}_s - \mathbf{x}^*\|^2 \\ &= -\|\nabla p(\mathbf{x}^*) - \hat{\mathbf{g}}_c(\mathbf{x}^*)\|\|\mathbf{x}_s - \mathbf{x}^*\| + \frac{\xi_1}{2}\|\mathbf{x}_s - \mathbf{x}^*\|^2 \\ &\geq -e_0\|\mathbf{x}_s - \mathbf{x}^*\| + \frac{\xi_1}{2}\|\mathbf{x}_s - \mathbf{x}^*\|^2 \end{aligned} \tag{7.25}$$

where ξ_1 is a positive number less than or equal to the least eigenvalue of the Hessian matrix of $p(\mathbf{x})$. For the function $p(r) = \frac{re^r}{k_4 H_0}$ and $r = \sqrt{\mathbf{x}^T\mathbf{x}}$, we have the following expression of the Hessian matrix of $p(\mathbf{x})$ according to the chain rule of Hessian matrices:

$$\begin{aligned} H(p(\mathbf{x})) &= \frac{d^2 p}{dl^2}\nabla l \nabla^T l + \frac{dp}{dl}H(l) \\ &= 4\frac{d^2 p}{dl^2}\mathbf{x}\mathbf{x}^T + 2\frac{dp}{dl}I \succeq 2\frac{dp}{dl} \\ &= \frac{e^r(\frac{1}{r}+1)}{k_4 H_0} > \frac{1}{k_4 H_0} \end{aligned} \tag{7.26}$$

where $l = r^2$, '$\succeq$' means the the eigenvalue of the left side is greater than the eigenvalue of the right side, i.e., the matrix equal to the left side minus the right side is semi-positive definite. Choosing $\xi_1 = \frac{1}{k_4 H_0}$, we have

$$p(\mathbf{x}_s) - p(\mathbf{x}^*) > -e_0\|\mathbf{x}_s - \mathbf{x}^*\| + \frac{1}{2k_4H_0}\|\mathbf{x}_s - \mathbf{x}^*\|^2 \tag{7.27}$$

Since $\mathbf{x}_s = \text{argmin}\{p(\mathbf{x})\}$, $p(\mathbf{x}_s) - p(\mathbf{x}^*) < 0$. Along with (7.27), we have,

$$-e_0\|\mathbf{x}_s - \mathbf{x}^*\| + \frac{1}{2k_4H_0}\|\mathbf{x}_s - \mathbf{x}^*\|^2 < 0 \tag{7.28}$$

i.e.,

$$\|\mathbf{x}_s - \mathbf{x}^*\| < 2k_4H_0e_0 \tag{7.29}$$

Step 3. From (7.29), if $2k_4H_0e_0 \leq r_2$,

$$\|\mathbf{x}_s - \mathbf{x}^*\| < r_2 \tag{7.30}$$

The inequality (7.30) indicates that the area centered at $\mathbf{x}_s$ with the radius r_2 is an region of attraction for the formation center, meaning that the robots will be attracted towards it when they are outside this region. Note that the replacement of $p(r)$ for $r < r_2$ with others has no influence to the convergence into the tumor area $r < r_2$. Therefore, the conclusion drawn above, which implies that the fact that the formation center is ultimately attracted to the tumor preserves for the true concentration function $p(r)$ for $r < r_2$.

Note that for a given r_2, we can choose the control parameter k_4, so that the center of the robot group converges to the tumor area with the radius r_2. In the situation where r_2 is not known prior, the last line of Algorithm 1 makes sure that the program stops when the robot group are in the tumor area with a pre-defined low pH value.

We claim theoretical convergence results in the following theorem.

Theorem 7.1 *Suppose* $\mathbf{g}_c(t)$ *is the true value of the gradient at the formation center, and its estimation* $\hat{\mathbf{g}}_c(t)$ *obtained by (7.12) has a bounded error, i.e.,* $\|\hat{\mathbf{g}}_c(t) - \mathbf{g}_c(t)\| \leq e_0$ *(*e_0 *is a positive constant). Using the control law (7.14), the relative positions of all nanorobots stabilize to the desired formation* $\bar{\mathbf{x}}_v$*, and the formation center* $\mathbf{x}_c(t) = \frac{1}{n}\sum_{i=1}^{n}\mathbf{x}_i(t)$ *converges to a point within the tumor region* $r < r_2$ *if we choose control parameters* k_4, c_0, c_2, *such that*

$$\begin{aligned} &2k_4H_0e_0 \leq r_2 \\ &\frac{c_2}{c_0} + \sqrt{k}e_0 < 0 \end{aligned} \tag{7.30}$$

The proof of the theorem is straightforward following the three-step analysis above.

In the next section, we present simulation results of the above control algorithm.

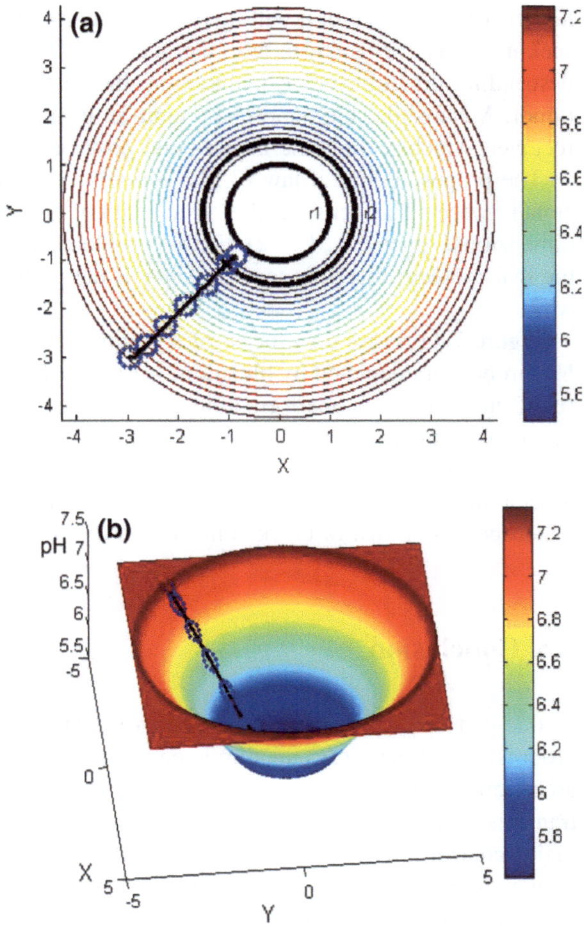

Fig. 7.4 **a** The process of a group of 12 nanorbots in tumor detection. **b** The process is depicted in a 3D plot. The *color bar* denotes the pH value defined in Eq. (7.6)

7.5.3 Simulation Results

Chemical detection in a complex dynamic environment is an important factor to consider for nanorobots interacting with the human body. For a solid tumor detection and therapy, the nanorobots are designed to circulate in the blood-stream and accumulate to the solid tumor. Eventually, the nanorobots release drug for treatment. In our study, given the model of tumor growth, a tumor pH diffusion environment is simulated, as shown in Fig. 7.3. These changes on acid concentration (pH value) are used to guide the nanorobots to locate the solid tumor in the human body.

The nanorobots may estimate gradients of the pH value. If the gradient is in low quantities, the seeking behavior of nanorobots will not be invoked. In such a case the nanorobots ignores the change of acid concentration assuming that it is within the expected levels of bloodstream. However, if the gradient reaches a value higher than a threshold, it will invoke the nanorobots' seeking behavior. When the seeking behavior gets started, the nanorobots move along the pH gradient-descending direction while maintaining the formation and eventually locate the tumor. Meanwhile, the nanorobots emit an electromagnetic signal detected by a receiver, which records the nanorobots' position where the signal happened.

The proposed control law (7.14) works for the problem defined in a k-dimensional space. To facilitate the visualization, we perform simulations in a 2-dimensional space (i.e., $k = 2$). For the problems defined in other dimensions, e.g., the real problem in a 3-dimensional space, the problem can be solved in a similar way.

Figure 7.4a shows the process of a group of 12 nanorobots for tumor detection. Nanorobots are depicted by blue dots. Based on the dimension of blood vessel, we set 12 nanorobots as one formation group. When the nanorobots are in the tumor microenvironment, they trace pH descending direction and locate the lowest pH area which is the position of the tumor. A 3D plot is used to show the tracing of pH descending gradient, as shown in Fig. 7.4b. The trajectory of the center of the nanorobots is shown in black solid line in the figure.

7.6 Conclusion

In this chapter, we propose a cooperative control strategy for pH sensitive nanorobots to deliver drugs in tumor environments. We first present a nanorobot architecture, including chemical sensors, actuators, power supply, and data transmission, which offers possible choices for manufacturing approaches, and major control interface requirements for designing medical nanorobots. We then review tumor pH measurement and microenvironment modeling. Based on a mathematical model of the ion diffusion process, we design control strategies for nanorobots to trace the gradient of the measured pH values, and reach the tumor center with the lowest pH value. Explicit gradient estimation and cooperative control laws are constructed, and convergence analysis is performed utilizing Lyapunov stability methods. The capability of the control law is illustrated through simulating a scenario of drug delivery by a group of nanorobots. Although the methodology introduced by the chapter is in the stage of conceptual design without materialized implementation, it is our intention to apply new robotics technologies to biomedical problems in order to create a solution that can be incrementally improved as the technology becomes available.

Acknowledgments Partial support for this work was provided by the National Science Foundation's Course, Curriculum, and Laboratory Improvement (CCLI) program under Award No. 0837584. Any opinions, findings, and conclusions or recommendations expressed in this material are those of the authors and do not necessarily reflect the views of the National Science Foundation.

References

1. Lamprecht A, Ubrich N, Yamamoto H, Schafer U, Takeuchi H, Maincent P, Kawashima Y, Lehr CM (2001) Biodegradable nanoparticles for targeted drug delivery in treatment of inflammatory bowel disease. J Pharm Exp Ther 299(2):775–781
2. Ummat A, Sharma G, Mavroidis C, Dubey A (2005) Biomedical engineering handbook, chapter Bio-nanorobotics: state of the art and future challenges. CRC Press, London
3. Cavalcanti A, Shirinzadeh B, Fukuda T, Ikeda S (2009) Nanorobot for brain aneurysm. Int J Robot Res 28(4):558–570
4. Cavalcanti A, Shirinzadeh B, Kretly L (2008) Medical nanorobotics for diabetes control. Nanomedicine 4(2):127–138
5. Cavalcanti A, Shirinzadeh B, Freitas R, Hogg T (2008) Nanorobot architecture for medical target identification. Nanotechnol 19(1):1–15
6. Cavalcanti A, Freitas R (2005) Nanorobotics control design: a collective behavior approach for medicine. IEEE Trans Nanobiosci 4(2):133–140
7. Cavalcanti A (2003) Assembly automation with evolutionary nanorobots and sensor-based control applied to nanomedicine. IEEE Trans Nanotechnol 2(2):82–87
8. Freitas R (2005) What is nanomedicine? Nanomed 1(1):2–9
9. Cavalcanti A, Freitas R (2002) Autonomous multi-robot sensor-based cooperation for nanomedicine. Int J Nonlinear Sci Numer Simul 3(4):743–746
10. Lewis M, Bekey G (1992) The behavioral self-organization of nanorobots using local rules. In: Proceedings of IEEE international conference on intelligent robots and systems, pp 1333–1338
11. Chandrasekaran S, Hougen D (2006) Swarm intelligence for cooperation of bio-nano robots using quorum sensing. In Bio micro and nanosystems conference, p 104, San Francisco, June 2006
12. Pawel P, Bermdez I Badia S, Bernardet U, Knsel P, Carlsson M, Gu J, Chanie E, Hansson BS, Tim C, Pearce, Verschure PFMJ (2006) An artificial moth: chemical source localization using a robot based neuronal model of moth optomotor anemotactic search. Auton Robots 20:197–213
13. Macnab RM, Koshland DE (1972) The gradient-sensing mechanism in bacterial chemotaxis. In: Proc National Academy of Sciences of the USA 69(9):2509–2512
14. Wadhams George H, Armitage Judith P (2004) Making sense of it all: bacterial chemotaxis. Nat Rev Mol Cell Biol 5(12):1024–1037
15. Freitas R (1999) Nanomedicine, volume I: basic capabilities. Landes Bioscience, Georgetown
16. Patolsky F, Lieber C (2005) Nanowire nanosensors. Mater Today 8(4):20–28
17. Cui Y, Wei Q, Park H, Lieber C (2001) Nanowire nanosensors for highly sensitive and selective detection of biological and chemical species. Science 293(5533):1289–1292
18. Hahm J, Lieber C (2004) Direct ultrasensitive electrical detection of dna and dna sequence variations using nanowire nanosensors. Nano Lett 4(1):51–54
19. Grosios K, Traxler P (2003) Tyrosine kinase targets in drug discovery. Drugs Futur 28(7):679
20. Kishimoto J, Spurr N, Liao M, Lizhi L, Emson P, Xu W (1992) Localization of brain nitric oxide synthase (nos) to human chromosome 12. Genomics 14(3):802–804
21. Wright E, Sampedro A, Hirayama B, Koepsell H, Gorboulev V, Osswald C (2005) Novel glucose sensor. US patent, 0267154
22. Che J, Cagin T, Goddard W (2000) Thermal conductivity of carbon nanotubes. Nanotechnology 11(7):65–69
23. Thostenson E, Ren Z, Chou T (2001) Advances in the science and technology of carbon nanotubes and their composites: a view. Compos Sci Technol 61(13):1899–1912
24. Thostenson E, Li C, Chou T (2005) Nanocomposites in context. Compos Sci Technol 65(3–4):491–516

25. Tombler T, Zhou C, Alexseyev L (2000) Reversible electromechanical characteristics of carbon nanotubes under local-probe manipulation. Nature 405:769–672
26. Yoon H, Xie JN, Abraham JK, Varadan VK, Ruffin PB (2006) Passive wireless sensors using electrical transition of carbon nanotube junctions in polymer matrix. Smart Mater Struct 15(1):s14–s20
27. Levitsky IA, Kanelos P, Euler WB (2004) Electromechanical actuation of composite material from carbon nanotubes and ionomeric polymer. J Chem Phys 121(2):1058–1065
28. Kim P, Lieber CM (1999) Nanotube nanotweezers. Science 286(5447):2148–2150
29. Wei BQ, Vajtai R, Ajayan PM (2001) Reliability and current carrying capacity of carbon nanotubes. Appl Phys Lett 79(8):1172–1174
30. Jenkner M, Tartagni M, Hierlemann A, Thewes R (2004) Cell-based cmos sensor and actuator arrays. IEEE J Solid-State Circuits 39(12):2431–2437
31. Astier Y, Bayley H, Howorka S (2005) Protein components for nanodevices. Curr Opin Chem Biol 9(4):576–584
32. Seeman N (2005) From genes to machines: Dna nanomechanical devices. Trends Biochem Sci 30(3):119–125
33. Sauer A, Stanacevic M, Cauwenberghs G, Thakor N (2005) Power harvesting and telemetry in cmos for implanted devices. IEEE Trans Circuits Syst 52(12):2605–2613
34. Mohseni P, Najafi K, Eliades S, Wang X (2005) Wireless multichnnel biopotential recording using an integrated fm telemetry circuit. IEEE Tans Neural Syst Rehabil Eng 13(3):263–271
35. Norris T (2006) Nanoacoustics: towards imaging nanostructures using picosecond ultrasonics. J Acoust Soc Am 119(5):3284–3285
36. Solberg JR, Lynch KM, Maciver MA (2008) Active electrolocation for underwater target localization. Int J Robot Res 27(5):529–548
37. Horiuchi TK, Cummings RE (2004) A time-series novelty detection chip for sonar. Int J Robot Autom 19(4):171–177
38. Ieee standard for safety levels with respects to human exposure to rf electromagnetic fields 3kHz–300GHz, 1999
39. Vaillancourt P, Djemouai A, Harvey J, Sawan M (1997) Em radiation behavior upon biological tissues in a radio-frequency power transfer link for a cortical visual implant. In: Proceedings of IEEE engineering in medicine and biology society, pp 2499–2502, Chicago
40. Engin K, Leeper D, Cater J, Thistlethwaite A, Tupchong L, Mcfarlane J (1995) Extracellular ph distribution in human tumors. Int J Hyperth 11(2):211–216
41. Oh K, Ohb Y, Ohc N, Lee KK, Lee E (2009) A smart flower-like polymeric micelle for ph-triggered anticancer drug release. Int J Pharm 375(1–2):163–169
42. Sethuraman V, Lee M, Bae Y (2008) A biodegradable ph-sensitive micelle system for targeting acidic solid tumors. Pharm Res 25(3):657–666
43. Wike-Hooley J, Haveman J, Reinhold H (1984) The relevance of tumor ph to the treatment of malignant disease. Radiother Oncol 2(4):343–366
44. Zhang X, Lin Y, Gillies R (2010) Tumor ph and its measurement. J Nucl Med 51(8):1167–1170
45. Liu G, Li Y, Pagel M (2007) A single paracest mri contrast agent for accurate in vivo ph measurements. In: International society for magnetic resonance in medicine, p 3406, Berlin, May 2007
46. Bhujwalla Z, Artemov D, Ballesteros P, Cerdan S, Gillies RJ, Solaiyappan M (2002) Combined vascular and extracellular ph imaging of solid tumors. NMR Biomed 15(2):114–119
47. Garcia-Martin ML, Herigault G, Remy C, Farion R, Ballesteros P, Coles JA, Cerdan S, Ziegler A (2001) Mapping extracellular ph in rat brain gliomas in vivo by 1h magnetic resonance spectroscopic imaging: comparison with maps of metabolites. Cancer Res 61(17):6524–6531
48. Gatenby R, Gawlinski E (1996) A reaction-diffusion model of cancer invasion. Cancer Res 56:5745–5753

49. Martin N, Gaffney E, Gatenby R, Maini P (2010) Tumour-stromal inter-actions in acid-mediated invasion: a mathematical model. J Theor Biol 267(3):461–470
50. Patel A, Gawlinski E, Lemieux S, Gatenby R (2001) A cellular automaton model of early tumor growth and invasion: the effects of native tissue vascularity and increased anaerobic tumor metabolism. J Theor Biol 213(3):315–331
51. Smallbone K, Gavaghanb D, Gatenbyc R, Mainia P (2005) The role of acidity in solid tumour growth and invasion. J Theor Biol 235(4):476–484
52. Smallbone K, Gatenby R, Maini P (2008) Mathematical modelling of tumour acidity. J Theor Biol 255(1):106–112
53. Li S, Guo Y (2012) Distributed source seeking by cooperative robots: all-to-all and limited communications. In: Proceedings of IEEE international conference on robotics and automation, pp 1107–1112, Saint Paul, May 2012
54. Slotine JJ,Li W (1990) Applied nonlinear control. Prentice Hall,Englewood Cliffs
55. Hassan Khalil (2002) Nonlinear systems,3rd edn, Prentice Hall

Chapter 8
Investigation of Protein–Protein Interactions in Cancer Targeted Therapy Using Nanorobots

Mi Li, Lianqing Liu, Ning Xi, Yuechao Wang, Zaili Dong, Xiubin Xiao and Weijing Zhang

Abstract This chapter describes the applications of nanorobots in investigating the mechanisms of rituximab's different efficacies in the targeted therapy of B-cell lymphomas at the individual cellular/molecular level. The chapter begins with an introduction to the new challenges in the field of cancer targeted therapy, taking rituximab targeted therapy in B-cell lymphoma for example. The following section presents a detailed description of the principles and methods of single-molecule techniques based on nanorobots. Next, it presents the microfabricated pillar-based cell immobilization method and discusses how to obtain the topography of individual living mammalian suspension cells based on this immobilization method. Next, it presents how to use nanorobot indentation experiments to measure the mechanical properties of individual cells. Next, the processes of using nanorobots to measure the individual molecular binding forces and three-dimensionally visualize the distribution of CD20 proteins on the lymphoma cell surface with the probe functionalization technology are detailed. The remainder of the chapter presents the specific binding force measurements on different lymphoma patients' cells and discusses its relation to rituximab's variable efficacies. The intent of this

M. Li · L. Liu (✉) · Y. Wang · Z. Dong
State Key Laboratory of Robotics, Shenyang Institute of Automation,
Chinese Academy of Sciences, Shenyang 110016, China
e-mail: lqliu@sia.cn

M. Li
e-mail: limi@sia.cn

N. Xi (✉)
Department of Electrical and Computer Engineering, Michigan State University,
East Lansing, MI 48824, USA
e-mail: xin@egr.msu.edu

X. Xiao · W. Zhang
Department of Lymphoma, Affiliated Hospital of Military
Medical Academy of Sciences, Beijing 100071, China

Y. Guo (ed.), *Selected Topics in Micro/Nano-Robotics for Biomedical Applications*,
DOI: 10.1007/978-1-4419-8411-1_8, © Springer Science+Business Media New York 2013

chapter is to provide the practical knowledge to begin the investigations on individual cells and molecules with nanorobots.

8.1 Background

Cancer has become the biggest killer of human health, one in four deaths in the United States is due to cancer [1]. The main treatments for cancer are surgery, radiotherapy, and chemotherapy. Surgery and radiotherapy are commonly used in the treatment of localized cancers, and in most cases are not sufficient to control metastatic cancers [2]. Chemotherapy arose in the 1940s and many decades have passed, but the therapeutic effects still do not fully meet our expectations [2]. Especially for advanced cancers, due to the loss of optimal surgery time, high-dose chemotherapy is often needed to sustain patients' survival time, during which the toxicity and side effects of chemotherapy bring great agony to patients and their families. In recent years, a novel therapy, called molecular targeted therapy, has been under the spotlight for its selectivity for cancer cells. Unlike chemotherapy drugs which kill both malignant and healthy cells, targeted drugs only kill the malignant cells. A targeted therapy drug can interfere with a specific molecular target (typically a protein) that is believed to have a critical role in tumor growth or progression [3]. For example, the target of Trastuzumab (a targeted drug for treating breast cancer) is human epidermal growth factor receptor 2 (HER2/neu) which is a cell membrane protein on breast cancer cells, the target of Bevacizumab (for colorectal, lung cancers) is vascular endothelial growth factor (VEGF) which is a signal protein produced by cells that stimulate angiogenesis, and the target of Rituximab (for B-cell lymphoma) is CD20 which is a transmembrane protein on lymphoma cells [4].

Rituximab is the first monoclonal antibody(mAb) targeted drug [5], which was approved by the US Food and Drug Administration (FDA) in 1997 for the treatment of B cell non-Hodgkin's lymphoma. Non-Hodgkin's lymphoma (NHL) is a disease in which cancer cells form in a person's lymphatic systems and start to grow uncontrollably. NHL constitutes 4 % of all tumors [1] and 85 % of NHL is of B cell origin [5]. Rituximab induces depletion of the target cell by binding with the CD20 on the cell surface through three mechanisms, antibody-dependent cellular cytotoxicity (ADCC), complement-dependent cytotoxicity (CDC) and programmed cell death (PCD) [6], as shown in Fig. 8.1. The approval of Rituximab revolutionized the treatment of B cell NHL. The use of Rituximab, particularly in combination with various chemotherapy/radiotherapy regimes, has significantly improved all aspects of the survival statistics for lymphoma patients [7]. Consequently, it has become the mainstream treatment for B-cell lymphoma [7].

For a little more than 10 years mAb has been a major aspect of clinical medicine: more than 25 antibodies are approved for human therapy and more than 240 antibodies are currently in clinical development worldwide [9]. Despite the unprecedented success of targeted drugs, there are still two big challenges for

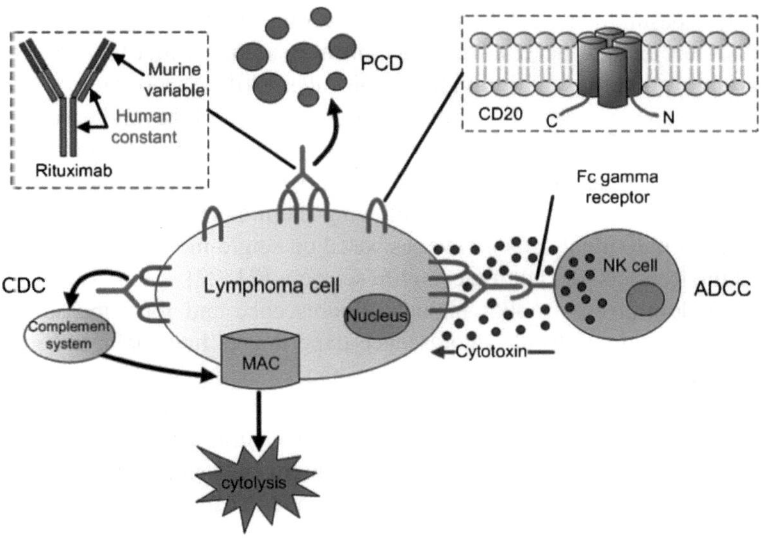

Fig. 8.1 The killing mechanism of rituximab. The binding of rituximab to CD20 on the lymphoma cell surface induces the depletion of the lymphoma cell through ADCC, CDC, and PCD. First, by ADCC, where rituximab binds to the Fc gamma receptors of natural killer (NK) cells, which triggers the release of cytotoxin to kill the targeted cell [8]. Second, by CDC, where rituximab activates the complement system and generates the membrane attack complex (MAC) on the targeted cell, which induces lysis of the targeted cell [4]. Third, by PCD, where Rituximab induces apoptosis of the targeted cell [6]

the further development of targeted therapy. One challenge is how to enhance the targeted drug's efficacy [10]. Virtually all targeted therapies are limited by the appearance of resistance to targeted drugs [11]. The clinical practice of molecular targeted therapy indicates that for the same treatment, different patients often express distinct responses. Some patients have benefited from targeted therapy, while others are unresponsive or develop resistance to the targeted drug [12]. The other challenge is to find biomarkers that can predict who will benefit from a particular targeted therapy [13] or that can be used for early cancer detection [14]. On the one hand, we urgently need cancer biomarkers that can indicate whether the outcome for the patient is likely to be good or poor and can also help doctors to decide which patients are likely to respond to a given drug and at what dose it might be most effective [13]. On the other hand, biomarkers that can be used for early detection is strongly desired, for the reason that cancers detected at advanced stages are far more likely to cause death than those detected while the cancer is still confined to the organ of origin [15].

Now virtually all knowledge of cellular chemical processes are deduced from in vitro ensemble measurements [16–18]. However, the results may not reflect the complete processes in vivo owing to their inherent ensemble averaging [17, 18]. The test tube measurements provide averaged information obtained on large ensembles of molecules from many cells [19]. Undoubtedly the results deduced from the ensemble measurements are correct for depicting the behavior of the ensemble molecules, but on the other hand the ensemble measurements hide the rare events and the variable properties of individual molecules [20]. As opposed to test tube experiments, single-molecule techniques allow one to look beyond ensemble averages [16] and thereby can reveal the events and properties that would otherwise be inaccessible, providing novel and complementary information for understanding the cellular chemical interactions and elucidating the molecular mechanisms. Hence, researches of individual cells/molecules based on single-molecule techniques have become the frontiers and hot topics in life sciences [21, 22].

The single-molecule toolkit includes fluorescence and force-measuring techniques [20]. Single-molecule fluorescence experiments have achieved great success in life sciences and improved our understanding of many fundamental biological processes [23, 24]. However, chemical labeling of proteins might change their surface characteristics so that their natural activity is impaired, and the varying labeling efficiency for different proteins makes accurately quantifying the detection difficult [25]. Hence, Label-free noninvasive assays are desired for dynamic monitoring of cellular status [26]. Compared with fluorescence techniques, single force-measuring techniques do not require the labeling of the target molecule [23], whereas they can exert forces on the molecules [20]. The main force-measuring techniques are optical tweezers, magnetic tweezers, and atomic force microscopy (AFM)-based nanorobots [27]. Optical tweezers use light to exert force on a transparent bead that is tethered to a surface by a single molecule, while magnetic tweezers use an external magnetic field to exert force on a superparamagnetic bead that is tethered to a surface by a single molecule [28]. The major drawback of optical tweezers is that it causes photodamage to the heating sample, while the drawback of magnetic tweezers is that it has no manipulation [27]. Compared with optical and magnetic tweezers, nanorobots is the only force probing technique that can simultaneously localize and manipulate single molecules on live cells [20]. Besides, nanorobots can work in various conditions, such as air, liquids, and vacuum, can visualize live cells and native biomolecules without the need for staining or fixation [29]. These particular advantages make nanorobots widely applied in life sciences [21, 22].

In this chapter, we describe the application of nanorobots in investigating the mechanisms of rituximab's different efficacies in the targeted therapy of B-cell lymphoma at the individual cellular/molecular level. The structure of this chapter is as follows. We begin by providing an introduction about nanorobots (Sect. 8.2.1) and a description of the single-molecule techniques based on nanorobots, including cellular stiffness measurements (Sect. 8 2.2), probe functionalization (Sect. 8.2.3), and molecular force and distribution measurements (Sect. 8.2.4). In the next section, we first discuss the cell immobilization methods for nanorobot cell imaging (Sect. 8.3.1). Then chemically fixed lymphoma cell imaging (Sect. 8.3.2) and living lymphoma cell imaging (Sect. 8.3.3) based on microfabricated pillars are presented. Next, the investigations of dynamically measuring the rituximab-induced changes of mechanical properties in B-lymphoma cells are presented (Sect. 8.4). In the next two sections, we discuss studies in which nanorobot force spectroscopy was used to measure the CD20-rituximab binding forces on isolated CD20 molecules, lymphoma Raji cells (Sect. 8.5), and on lymphoma patient cells (Sect. 8.6), respectively. The last section is the summary (Sect. 8.7).

8.2 Single-Molecule Techniques Based on Nanorobot: Principles and Methods

8.2.1 An Introduction to Nanorobots

Nanorobotics is a recently emerged and rapidly developed area in systems and controls research field [30, 31]. Nanorobotics can access the unprecedented and small areas [30] to obtain information at the nanoscale which is inaccessible for macroscopic robots. Nanorobotics includes robots that are nanoscale in size called nanorobots, and large robots capable of manipulating objects that have dimensions in the nanoscale range with nanometer resolution, called nanorobotic manipulators [32]. For nanorobots, they are naturally related to medicine, being strongly expected to improve the quality of human health. We can imagine that a doctor injects a special type of nanorobot into the patient; these robots can arrive at the focus of the disease and then destroy the malignant cells, leaving healthy cells untouched [33]. In recent years, some designs of nanorobots have been presented. Freitas [34] presented a design of artificial mechanical red blood cell which is able to deliver 236 times more oxygen to the tissues per unit volume than natural red blood cells and to manage carbonic acidity. Cavalcanti et al. [35] designed a nanorobot for diabetes control which can flow through the vessel with the bloodstream to pervasively monitor the glucose levels. These nanorobots are fascinating and may bring revolutionary changes in human life. However, nanorobot is still a fantasy due to the lack of micromanufacturing technology which can fabricate nanorobot to its final form [36].

Compared with nanorobot, nanorobotic manipulator has achieved significant development since the invention of AFM [37] in 1986. AFM is the main component of a nanorobotic manipulator. The principle of AFM is shown in Fig. 8.2. AFM is mainly composed of probe, piezoelectric actuator, position sensitivity detector (PSD), and signal processing & feedback loop. AFM renders the surface topography images of the sample by raster scanning a tip attached to the end of a cantilever over the sample surface, while monitoring the changes in cantilever deflection with a split photodiode PSD. The feedback loop maintains a constant deflection between the cantilever and the sample by vertically moving the actuator to maintain a "setpoint" deflection to keep the interaction forces between the atoms on the tip and the atoms on the sample constant.

Combined with teleoperation, AFM can be used to manipulate nanoobjects with the aid of augmented reality and haptic feedback technology [38], by pushing,

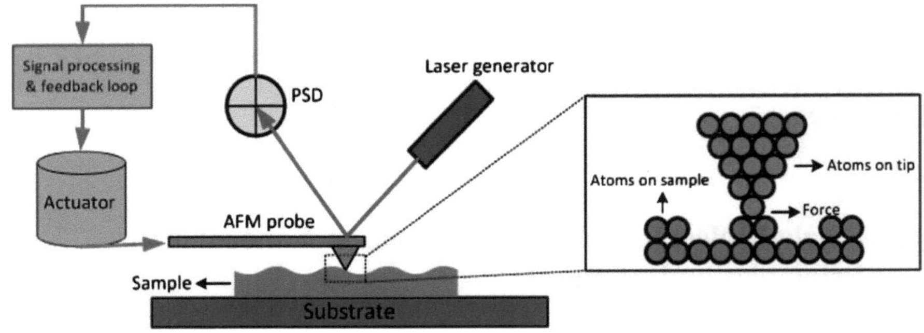

Fig. 8.2 Principle of AFM

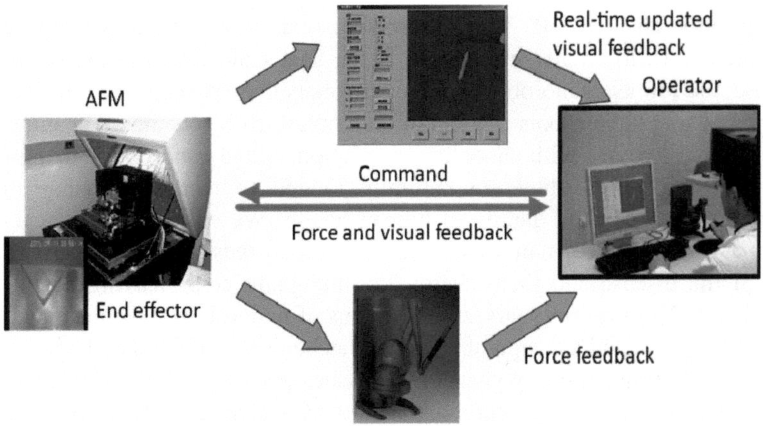

Fig. 8.3 Nanorobotic manipulator based on AFM

pulling, cutting, picking & placing, positioning, orienting, assembling, indenting, bending, and twisting nanoobjects [39]. As shown in Fig. 8.3, a human operator directly in the control-loop manipulates the nanoobjects by using a man–machine interface [40]. First, an AFM image is produced by scanning the nanoobjects. This image is input into the man–machine interface as the reference scene. Second, the operator manipulates the nanoobjects through the joystick, during which the reference scene on the man–machine interface real-time updates. The operator adjusts the manipulation according to the visual feedback of the reference scene and the force feedback of the joystick until the task is finished. For convenience, nanorobot is referred to AFM-based nanorobotic manipulator in the following sections.

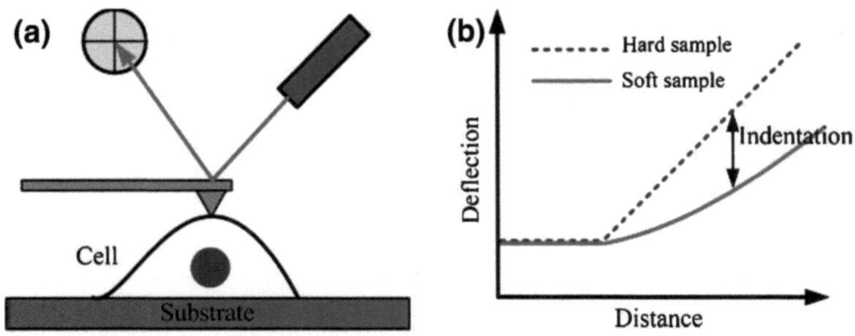

Fig. 8.4 Principle of measuring the cell's mechanical properties with nanorobot. **a** Obtaining *force curves* on cell surface. **b** *Force curves* obtained on hard sample and soft sample

8.2.2 Cellular Stiffness Measurement

The principle of measuring the cell's mechanical properties is shown in Fig. 8.4. At a fixed lateral position, the nanorobot probe is first approached and then retracted from the cell surface. The deflection of the probe cantilever and the displacement of the piezoelectric actuator are recorded during the approach–retract cycle. Then the force–distance curve is obtained, which is also called force curve. The force curve reflects the mechanical properties of the sample. If the force curve is obtained on hard sample (such as glass), the cantilever deflection is equal to the piezoelectric actuator movement, because there is no deformation on the sample. And therefore the force curve is straight. If the force curve is obtained on soft sample (such as cell), the cantilever deflection decreases, because the probe tip indents into the sample. Therefore the force curve is bent. Hence the force curve on the sample reflects its mechanical properties and the Young's modulus of the sample can be computed from force curve.

The most popular model used for describing the elastic response of a sample indented by a nanorobot tip is the Hertz model. The Hertz model is in the following equation [41]:

$$F_{\text{sphere}} = \frac{4ER^{1/2}\delta^{3/2}}{3(1-\upsilon^2)} \tag{8.1}$$

$$F_{\text{cone}} = \frac{2E\delta^2 \tan\theta}{\pi(1-\upsilon^2)} \tag{8.2}$$

where υ is the Poisson ratio of the sample (often assuming the Poisson ratio of the cell is 0.5 [41]), δ is the indentation depth, R is the radius curvature of the tip, θ is the half-opening angle of the nanorobot tip, E is the Young's modulus, and F is the applied loading force.

According to Hook's law, the loading force can be computed.

$$F = kx \tag{8.3}$$

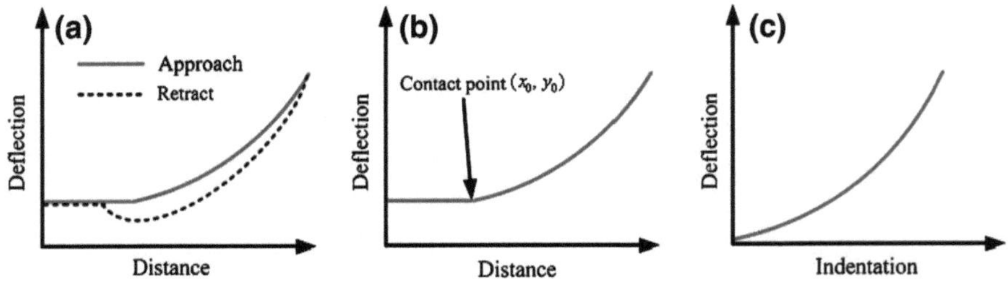

Fig. 8.5 Process of converting *force curves* into *indentation curves*. **a** Original *force curve*. **b** *Approach curve* with contact point (x_0, y_0). **c** *Indentation curve*

where k is the spring constant of the probe and x is the deflection of the cantilever.

The process of computing the cell's Young's modulus from the force curve is as shown in Fig. 8.5. Each force curve contains two portions, approach curve and retract curve, as shown in Fig. 8.5a. Each force curve responds to a nanorobot probe's approach–retract cycle on cell surface. The lateral coordinate of force curves is the displacement of the piezoelectric actuator and the vertical coordinate is the cantilever deflection. Because the retract curve is often influenced by adhesion force, approach curve is selected to compute the Young's modulus. The force curve is converted into the indentation curve according to the contact point(x_0, y_0) on the approach curve (Fig. 8.5b). The convert process is followed thus. ① Defining the coordinate of the point on the approach curve is (x, y), x is lateral coordinate, y is vertical coordinate. ② Count the points whose lateral coordinate x is smaller than the contact point's lateral coordinate x_0. Then sum the vertical coordinate of these points and compute the mean value y_1. Subtract y_1 from the vertical coordinate y for all the points on the approach curve and new coordinate(x, y_2), $y_2 = y - y_1$ of these points is obtained. ③ Count the points whose lateral coordinate x is bigger than the contact point's lateral coordinate x_0. Subtract the sum of the x_0 and y_2 from x and new coordinate(x_2, y_2), $x_2 = x - (x_0 + y_2)$ of these points is obtained. ④ Plot the (x_2, y_2), and indentation curve is constructed (Fig. 8.5c).

8.2.3 Probe Functionalization

The prerequisite of SMFS experiments is linking specific proteins onto the surface nanorobot tip, which is called probe functionalization. There are now three methods for probe functionalization, biotin–avidin method, NTA-histidine method, and polyethylene glycol (PEG) method [42]. In 1994, Florin et al. [43] measured the binding force of biotin–avidin on biotinylated agarose bead, by linking avidin to nanorobot tip. Then in 1998, Fritz et al. [44] measured the binding force of P-selectin and P-selectin glycoprotein ligand-1. In this method, the nanorobot tip is first coated with biotinylated BSA, and then avidin is bound to

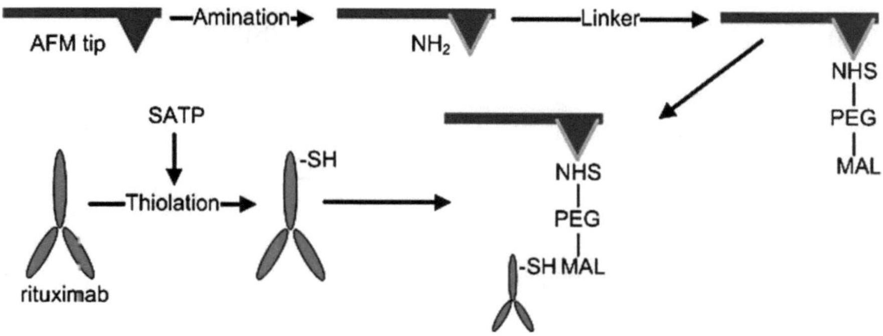

Fig. 8.6 Process of linking rituximab onto the nanorobot tip

the biotin on the tip. The protein is functionalized with biotin and then linked onto the tip by binding with avidin. The second method is NTA-histidine [45]. In this method, the nanorobot tip is coated with a 5-nm thick chromium layer followed by a 30-nm thick gold layer. Then NTA-terminated and tri [ethylene glycol (EG)]-terminated alkanethiols is linked onto the nanorobot tip. Finally the protein molecule is functionalized with histidine which can bind the Ni^{2+}-NTA group. The first two methods require that the tip is coated with a gold layer. The third method can be performed directly on silicon tip, by PEG linker [46, 47]. The most common PEG linker is NHS-PEG-PDP. The linker has an NHS ester on one end and a PDP ester on the other. The tip is coated with amino groups which can bind the NHS ester of the PEG. The protein is treated to form thiol groups which can bind the PDP ester of the PEG.

To measure the binding force of individual CD20-rituximab, we need first to link rituximab molecules onto the surface of the nanorobot tip. The heterobifunctional linker NHS-PEG2000-MAL was used to tether rituximabs onto the nanorobot tip. This linker has an NHS ester on one end and a MAL ester on the other. The tip functionalization process followed an established procedure [47], as shown in Fig. 8.6. ① Use 30 μL APTES and 10 μL NN-diisopropylethylamine to coat the nanorobot tip with amino groups (NH_2) in a glass dessicator under argon for 0.5–2 h. ② Mix the PEG linker and triethylamine in chloroform and then incubate the NH_2-modified probes in this solution for 2–3 h. ③ Treat rituximabs with SATP to form thiol functions. The thiols can react with the MAL end of the PEG linker [48]. ④ Finally put the probes in the mixture of SATP-rituximab, hydroxylamine and buffer solution (pH 7.5) for 1 h. The functionalized probes were stored in PBS at 4 °C.

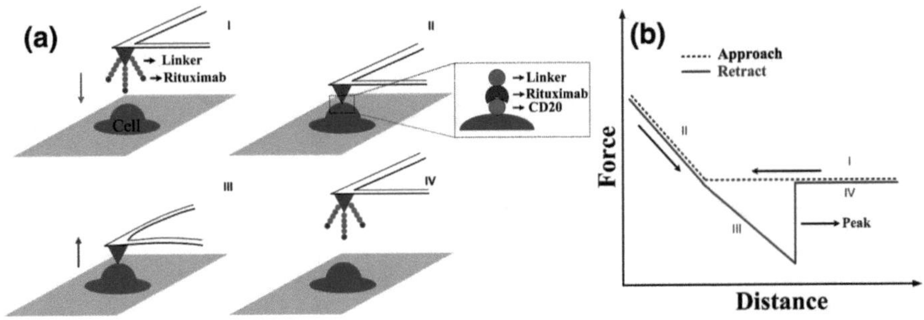

Fig. 8.7 Principle of SMFS. **a** Functionalized probe is used to obtain *force curve* on the cell surface. **b** *Force curve*

8.2.4 Molecular Force and Distribution Measurement

The nanorobot can work in fluids, air, and vacuum, with nanometer positional accuracy and piconewton force sensitivity [42]. This unique advantage makes it suitable to detect forces between single receptor-ligand pairs, and this technology is termed single-molecule force spectroscopy (SMFS) [49]. The principle of SMFS is shown in Fig. 8.7. Through probe functionalization technology, ligands are tethered onto the nanorobot tip surface by linker molecules. Then the functionalized probe carrying ligands is first approached and then retracted from the cell surface, which is obtaining force curves. Each approach–retract cycle can be divided into four stages. First, the tip descends and there is no contact with cell surface yet. No bending of the cantilever occurs at this stage. Second, the tip touches the cell surface, and the attractive forces pull the tip down. As the tip presses into the surface, cantilever bends upward due to the repulsive tip-sample interaction. Third, after the deflection of the cantilever reaches its maximum value, the tip begins to retract from cell surface. When ligand binds the receptor, the receptor-ligand complex is stretched and the attractive force from the stretching complex causes the cantilever to bend downwards (Fig. 8.7b). When the retract force exerted by the cantilever is equal to the complex binding force, the complex ruptures and the cantilever returns to its original position. Hence, if receptor-ligand binding occurs during the approach–retract cycle, there is an abrupt peak on the retract portion of the force curve.

The functionalized tip which carries proteins can also be used to measure the distribution density of the cognate specific proteins on cell membrane. Now there are two methods for measuring the distribution of a particular protein on cell membrane. The first method is to obtain an array of force curves (16 × 16) in a local area with the functionalized probe [45]. For each force curve, the binding force is computed. Then these binding forces are mapped into a two-dimensional gray image. The gray map reflects the particular protein distribution on cell surface. The second method is to directly scan the local area on the cell surface with the functionalized probe at tapping mode [50]. The phase image is filtered by a

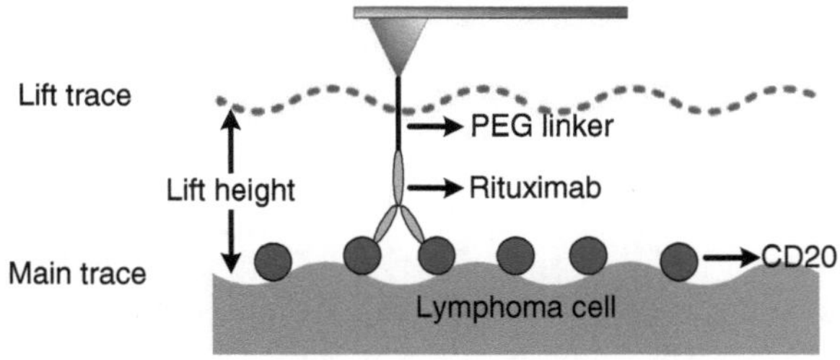

Fig. 8.8 Principle of lift scan mode to measure the CD20 distribution

band-pass filter to remove the low-frequency topographical information and the high-frequency noise. The distribution of the particular protein is then visualized in three-dimensional phase image.

To visualize the distribution of CD20 molecules on the lymphoma cell surface, nanorobot lift scan mode is used here, as shown in Fig. 8.8. The functionalized probe which was linked with rituximab molecules was used to perform the lift scan. In this mode, the main trace is first made to typically measure the topography, and then the lift trace is made along the main trace to produce an image of the CD20-Rituximab-specific recognition. Because the disturbance from the topography has been eliminated, the lift trace is then only influenced by the CD20-Rituximab binding. Consequently, the location of CD20 on the cell surface can be extracted. The lift scan height is the distance between the main trace and the lift trace and approximates the length of the PEG linker molecule.

8.3 Cell Immobilization: The Prerequisite for Nanorobot Investigation

8.3.1 An Introduction to Cell Imaging with Nanorobot

The basic capability of a nanorobot is to obtain the topography of the sample and the nanorobot was first applied in the early 1990s to cell imaging [51]. After about 20 years of development, using nanorobots to image cells has achieved great progress. However, imaging cells in fluids, particularly for living mammalian suspension cells, is still severely challenging. First, due to the softness of the cell membrane, the normal force exerted by the scanning tip considerably deforms the cell and often "soft-and-sticky" interactions between the tip and cellular surface exist [52]. Owing to this influence, so far individual protein molecules on living cells cannot be visualized by nanorobots (the imaging resolution on mammalian cells is limited to ~50 nm [21]). Second, the prerequisite of imaging cells in fluids is immobilizing the cells onto the substrate in order to avoid the displacement of cells caused by the scanning nanorobot probe [53]. According to the various types of cells, there are now three main methods for cell immobilization. The first method is to culture adherent cells on the substrate (such as glass slide) for about

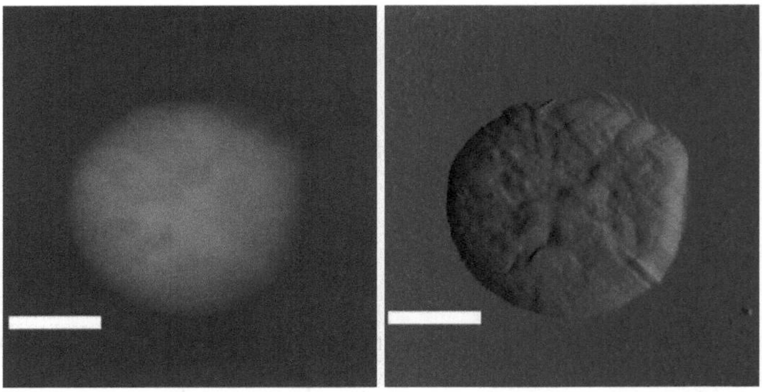

Fig. 8.9 Chemical fixed lymphoma Raji cell [61]. Topography image (*left panel*). Deflection image (*right panel*). Scale bar is 10 μm

1–2 days and these cells can adhere to the substrate tightly [53]. The second method is to cover the substrate with positively charged macromolecules (such as poly-L-lysine [54, 55]) and cells can be attached on the substrate by electrostatic attraction interactions. The third method is to mechanically trap the cells in the pores with porous polymer membranes [56, 57] and this method is generally suited for microbial cells which cannot spread over the substrate [57]. For mammalian suspension cells, Rosenbluth et al. [58] used microfabricated wells to mechanically immobilize leukocytes, Ng et al. [59] used pyramidal wells to immobilize chondrocytes, but both the methods are not convenient for nanorobot imaging. Through these immobilization methods mentioned above, hardy cells [60] like adherent cells and microbial cells can be imaged livingly by nanorobots. However, for mammalian suspension cells, immobilization methods that are simple and effective are still urgently needed.

8.3.2 Fixed Lymphoma Cell Imaging

To investigate Rituximab's mechanisms with nanorobots at cellular and molecular levels, the prerequisite is to image the individual lymphoma cells. One of the most common methods for cells' nanorobot imaging is by chemical fixation. We used nanorobots to image the B-lymphoma Raji cells after chemical fixation. The Burkitt's lymphoma Raji cells were cultured at 37 °C (5 %CO_2) in RPMI-1640 culture medium containing 10 % fetal bovine serum. Raji cells were attached on the glass slide by poly-L-lysine's electrostatic adsorption and glutaraldehyde's chemical fixation. The process of sample preparation is as follows. ① Cover the glass slide with poly-L-lysine and store the slide at room temperature overnight. ② Harvest cells by centrifuging at the speed of 1,000 rpm. ③ Drop the cell suspension onto the glass slide. ④ After 30 min, fix the cells with 0.25 % glutaraldehyde for 15 min and then put the slide onto a petri dish with PBS for nanorobot imaging experiments. The nanorobot images of Raji cells fixed by glutaraldehyde

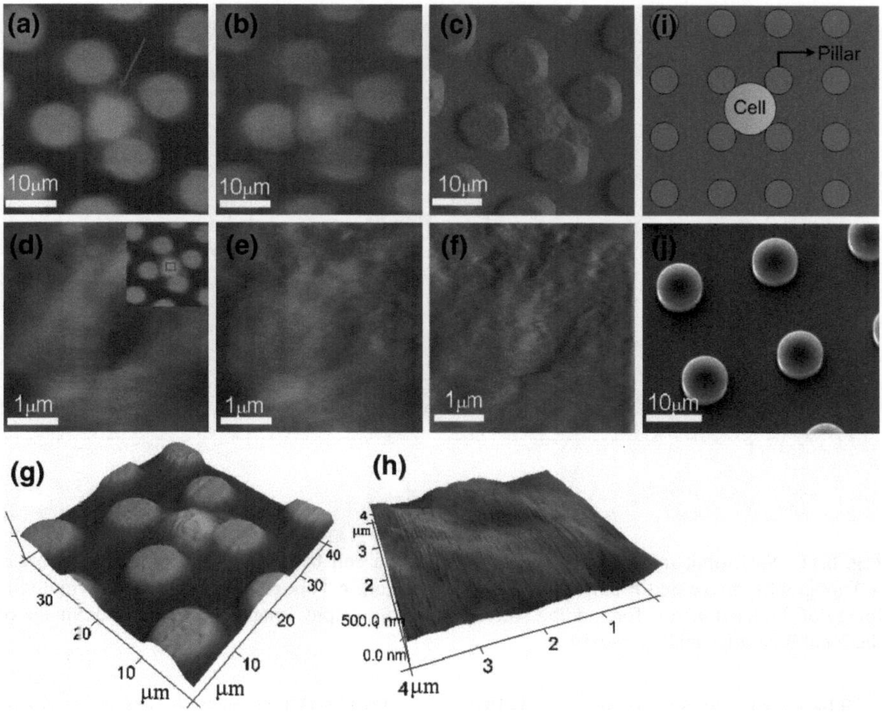

Fig. 8.10 Tapping mode images of lymphoma living cells trapped in microfabricated patterned pillars [62]. Height image (**a**), phase image (**b**), amplitude image (**c**) and three-dimensional height image (**g**) of the lymphoma cell trapped in the pillars (denoted by the *red arrow* in **a**). Higher resolution images were acquired by zooming into the local area of cell surface (denoted by the *red rectangle* in the *inset* in **d**), height image (**d**), phase image (**e**), amplitude image (**f**) and three-dimensional height image (**h**). Schematic diagram (**i**) and SEM image (**j**) of the pillars

are shown in Fig. 8.9. The images were obtained at contact mode with a Dimension 3,100 AFM (Veeco company). The scan rate was 1 Hz, and the scan size is 40 μm.

8.3.3 Living Lymphoma Cell Imaging

For living lymphoma cell imaging, we have developed a pillar-based immobilization method [62]. The height of the microfabricated patterned pillars is 5 μm, and the diameter of the pillars is 10 μm. The cell is trapped by four pillars for being imaged by nanorobots(Fig. 8.10i). The pillars are fabricated on a silicon substrate (the height is 525 μm and the diameter is 100 mm) by photolithography. The SEM image of the pillars was shown in Fig. 8.10j.

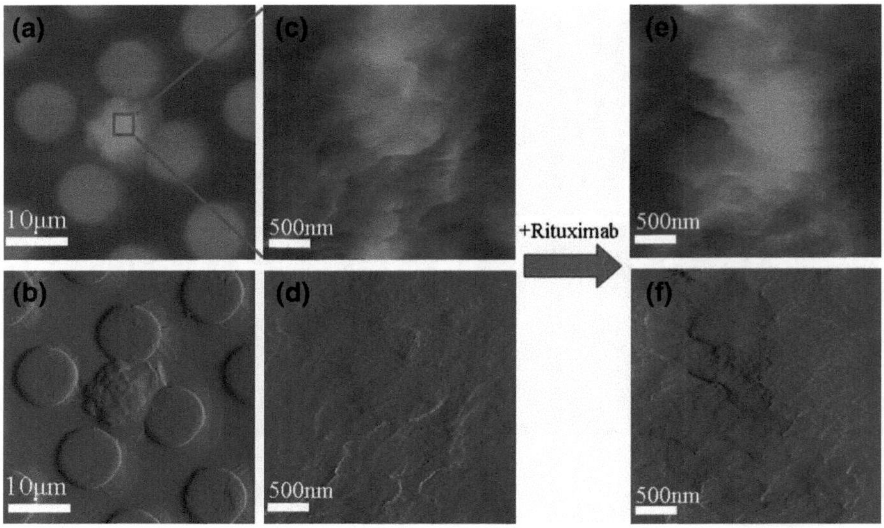

Fig. 8.11 Nanorobot images of a living B-lymphoma cell before and after adding rituximab. **a** Topography image and **b** amplitude image of the cell. **c** Topography image and **d** amplitude image of the local area before adding rituximab. **e** Topography image and **f** amplitude image of the local area after adding rituximab

The sample preparation is as follows: (1) Dilute 0.1 % poly-L-lysine 10 times with deionized water and then drop the poly-L-lysine dilution onto the pillars and stored for natural drying. (2) Add 5 ml of cell suspension into an eppendorf tube and centrifuge 5 min at the speed of 1,000 rpm. (3) Remove the supernatant and add fresh PBS to stir the solution and drop the solution onto the pillars. (4) Attach the pillars to a glass slide using a small piece of double-sided adhesive tape and then put the glass slide into a Petri dish in PBS for being imaged by nanorobots.

With the mechanical trapping of pillars, the topography image, phase image, and amplitude image of a living B-lymphoma cell are shown in Fig. 8.10a–c, respectively. The three-dimensional topography image is shown in Fig. 8.10g. We can see that a cell (denoted by the red arrow in the inset in Fig. 8.10a) was trapped in the middle of the four pillars. To image the topography of the local area of the cell surface, the scan size was zoomed into 4 μm (denoted by the red rectangle in Fig. 8.10d). The topography image, phase image, and amplitude image are shown in Fig. 8.10d–f, respectively. The three-dimensional topography image is shown in Fig. 8.10h. We can see the corrugated morphology of the cell surface and the results demonstrate that the pillar-based immobilization method is reliable and effective.

To visualize the topography changes of B-lymphoma cells stimulated by rituximab, one cell was chosen to be imaged before and after adding rituximab. Figure 8.11 showed the topography changes of a B-lymphoma cell after adding rituximab. Figure 8.11a was the topography image of the cell, and Fig. 8.11b was the amplitude image of the cell. Figure 8.11c is the topography image of the local

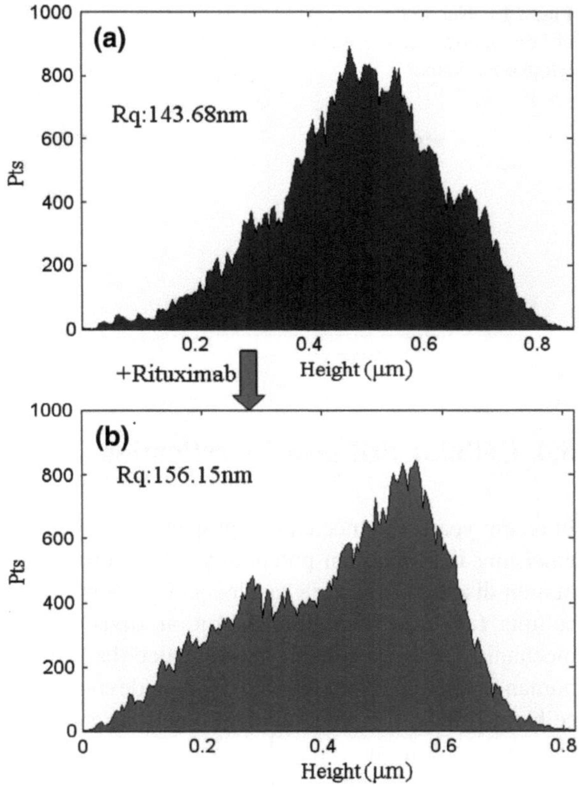

Fig. 8.12 Roughness of the local area topography images before and after adding rituximab. **a** Before adding rituximab. **b** After adding rituximab

area of the cell surface, and Fig. 8.11d is the amplitude image of the local area of the cell surface. After adding rituximab, the topography image of the local area of the cell surface is shown in Fig. 8.11e, and the amplitude image of the local area of the cell surface is shown in Fig. 8.11f. We can see that after adding rituximab, the cell surface was more corrugated. Figure 8.12 is the contrast of roughness of the local area topography images before and after adding rituximab. We can see that the roughness increased from 143.68 to 156.15 nm. This indicated that the stimulation of rituximab results in a rougher cell surface.

In summary, a platform based on microfabricated pillars was established for the investigation of individual living lymphoma cells with nanorobots. For the future work, we would like to separate living lymphoma patient tumor cells from normal cells (Fig. 8.13) and then investigate their behaviors in order to explore the patients' different rituximab resistance.

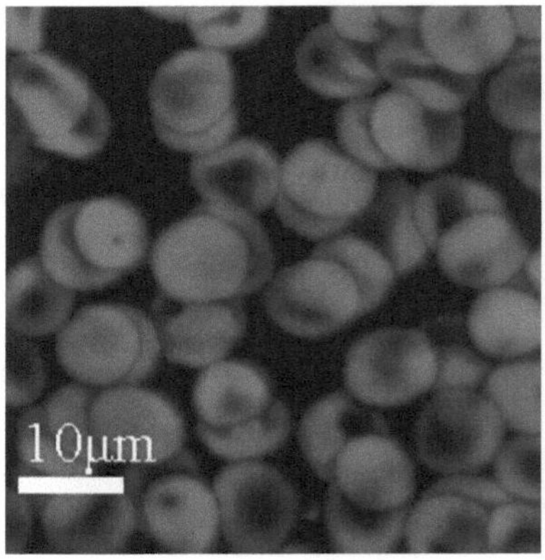

Fig. 8.13 Nanorobot images of bone marrow cells from a lymphoma patient

8.4 Cellular Stiffness Investigation

In recent years, the mechanical properties of single cells have been identified as an emerging field that can potentially make significant contributions in the study of human diseases [63]. Cell mechanical properties play an important role in critical cellular functions, including migration, division, and shape [29]. Studies into the mechanics of single cells have implicated that cell mechanics are closely related to human health and disease [64, 65]. Any deviations in the mechanical properties of cells may affect the physiological functions and give rise to disease, while disease can also result in mechanical property changes in living cells [66]. Hence studying human diseases from a biomechanical perspective can lead to a better understanding of the pathophysiology and pathogenesis of human diseases [63].

In the past decade, the nanorobot has been proven to be a powerful, versatile, and easy-to-use tool for investigating the mechanical properties of single cells [41, 58, 62–73]. First, nanorobots can work in fluids and this enables researchers to perform experiments on living cells in situ. Second, the nanometer resolution of nanorobots opens the possibility for measuring locally the mechanics of a cell and to distinguish different regions of the cell [73]. Third, the nanorobot is easy to control and the sample preparation is simple. With the nanorobot nanoindentation technique, the mechanical properties of various cells have been measured, including breast cancer cells [66], fibroblasts [67, 68, 72], microbial cells [41], leukemia cells [58, 69], normal rat kidney (NRK) cell [70], and lymphocytes [71], providing novel information for understanding the cellular physiological activities.

We measured the rituximab-induced changes of mechanical properties in B-lymphoma cells. First, AO/EB double fluorescent staining was used to detect the effect of B-lymphoma cell after the stimulation of rituximab. Briefly, 3×10^3 cells were resuspended in 1.5 ml culture medium and plated in 24-well flat-bottom microtiter plates. After incubating at 37 °C (5 %CO_2) for 24 h with rituximab (10 and 100 μg/ml), the cells were centrifuged at 1,000 rpm for 5 min and harvested onto glass slides. The AO/EB staining solution (100 μg/ml) was dropped onto the glass slides and incubated for 2 min. Then the glass slides were put onto the stage

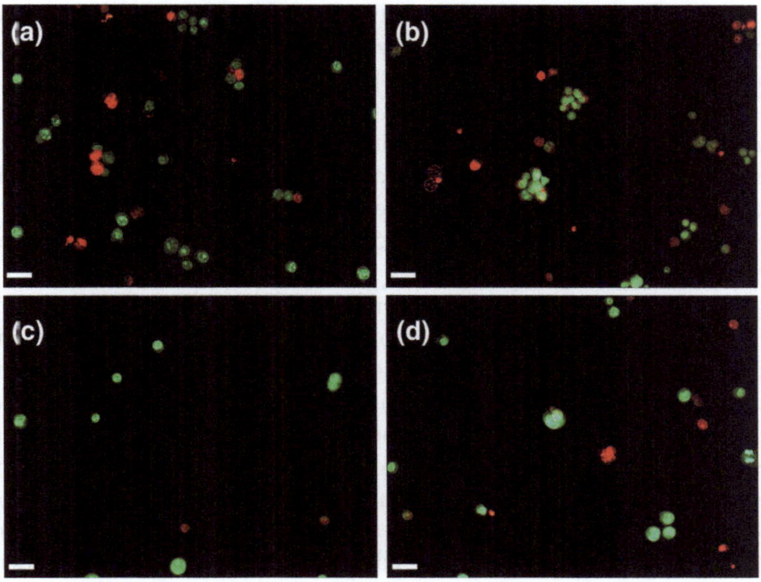

Fig. 8.14 Dual fluorescence merge images of detecting the effect of rituximab on B-lymphoma cell growth by AO/EB double fluorescent staining [62]. **a** 10 μg/ml rituximab. **b** 100 μg/ml rituximab. **c** Control experiments of **a**. **d** Control experiment of **b**. Scale bar is 20 μm

of the fluorescence microscope to acquire fluorescence images. For control experiments, cells that cultured without rituximab were detected by AO/EB double fluorescent staining. Then, the nanorobot was used to quantitatively measure the dynamic changes of mechanical properties of B-lymphoma cells.

AO/EB double fluorescent staining was used to detect the effect of rituximab on B-lymphoma cell growth, as shown in Fig. 8.14. The dual fluorescence merge images of AO/EB staining of B-lymphoma cells with 10 and 100 μg/ml rituximab are shown in Fig. 8.14a and b, respectively. The dual fluorescence merge images of AO/EB staining of B-lymphoma cells without rituximab are shown in Fig. 8.14c and d. From Fig. 8.14a we can see the distinct apoptosis of B-lymphoma cells, as the appearance of nuclear pyknosis and condensed chromatin at the periphery of the nuclear envelope. While for the cells cultured with 100 μg/ml rituximab, the apoptosis was not distinct. From the control experiments (Fig. 8.14c and d), the apoptosis was not observed. This indicated that the apoptosis was induced by rituximab. Rituximab has been shown to induce apoptosis in vitro [6]. Flieger et al. [74] have shown that rituximab alone induced apoptosis in Raji lymphoma cell lines at a concentration above 100 ng/ml. There is general agreement that hyper-cross-linking enhances the apoptotic effects of CD20 mAbs [75]. Shan et al. [76] have shown that the anti-CD20 mAb could induce the apoptosis in Ramos lymphoma cell lines in the presence of GAM cross-linker. The support for rituximab promoting cell death is apparent, but whether this mechanism is critical for the depletion of CD20-positive target cells in vivo remains to be proven [7]. Better understanding the potential interplay between the multiple proposed mechanisms will be of great significance for improving the efficacy of anti-CD20 mAb and decreasing the associated adverse events [77].

For measuring the rituximab-induced dynamic changes of mechanical properties in B-lymphoma cell quantitatively, rituximab(10 μg/ml) was added to the Petri dish and incubated for 1 h. The probe was localized onto the cell surface with the assistance of a CCD camera. Then force curves on the cell were obtained every 5 min at the same position. Each time 25 force curves were obtained at the same loading rate.

Figure 8.15 was the mechanical properties measurement results of B-lymphoma cells. The nanorobot deflection image of B-lymphoma cell without rituximab is shown in Fig. 8.15a. The deflection image of B-lymphoma cell with rituximab is shown in Fig. 8.15b. The force curves without rituximab are obtained at the center of the black square in Fig. 8.15a. The force curves with rituximab are obtained at the center of the red square in Fig. 8.15b. The typical force curves (only approach curves are shown) without (black line) and with (red line) rituximab are shown in Fig. 8.15c. We can see that the slope of the force curve with rituximab is smaller than that without rituximab and this indicates that the cells became softer after adding rituximab. The process of measuring the mechanical properties of B-lymphoma cells at a certain time is shown in Fig. 8.15d–e. Twenty-five force curves were obtained at the time. Of these 25 force curves, one typical force curve (only the approach curve is shown) is shown in the inset in Fig. 8.15d (denoted by the blue dotted line). The applied force is equal to the product of the deflection and constant spring of the cantilever. Then after converting the force curve into the indentation curve (the indentation depth is computed by subtracting the deflection of the cantilever from the Z movement), the Young's modulus of the cell can be computed by putting the value of indentation depth and applied force into the Hertz model formula. Several hundred values of Young's modulus were computed from the force curve and the histogram of these values are shown in the inset in Fig. 8.15d. The Gaussian fitting indicated that the Young's modulus from the force curve was 0.6933 ± 0.01123 MPa. The theoretical deflection was computed by putting the Young's modulus and indentation depth to the Hertz model formula. The fitting indentation curve is shown in Fig. 8.15d. From the comparison of the experimental indentation curve and fitting indentation curve, we can see that the two curves are consistent with each other. For the other 24 force curves, the process of computing the Young's modulus was the same as the above. The histogram of the 25 values of Young's modulus is shown in Fig. 8.15e. The Gaussian fitting indicated that the Young's modulus of B-lymphoma cell at the certain time was 0.6642 ± 0.03699 MPa.

The computation processes of the Young's modulus in other times were the same as described in Fig. 8.15d and e. The dynamic changes of mechanical properties without rituximab and with rituximab are shown in Fig. 8.15f and g, respectively. The horizontal coordinate is time and the vertical coordinate is Young's modulus. From Fig. 8.15f, we can see that the Young's modulus of B-lymphoma cells without rituximab increased from 0.2883 ± 0.0117 MPa

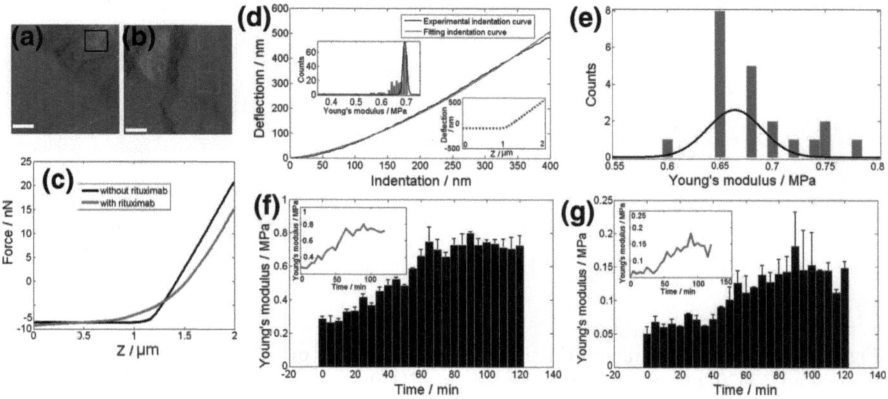

Fig. 8.15 Measuring the changes in mechanical properties of B-lymphoma cells without and with rituximab using nanorobots [62]. **a** Nanorobot deflection image of B-lymphoma cell without rituximab. **b** Nanorobot deflection image of B-lymphoma cell with rituximab. **c** Typical *force curves* (*approach curves*) without (*black line*) and with (*red line*) rituximab. The process of computing the Young's modulus of B-lymphoma cells at a certain time is shown in **d** and **e**. **d** Comparison of experimental indentation curve and fitting indentation curve. The Young's modulus was computed from a typical approach curve (denoted by *blue dotted line*). Histogram of several hundred values of Young's modulus computing from the approach curve is shown in the inset. **e** Histogram of 25 values of Young's modulus from 25 force curves obtained at the certain time. **f** Real-time changes of mechanical properties of B-lymphoma cell without rituximab. **g** Real-time changes of mechanical properties of B-lymphoma cell with rituximab. Scale bar is 2 μm

(0 min) to 0.7259 ± 0.05853 MPa (120 min) in 2 h. From Fig. 8.15g, we can see that the Young's modulus of B-lymphoma cells with rituximab increased from 0.05045 ± 0.0104 MPa (0 min) to 0.1491 ± 0.008901 MPa (120 min) in 2 h. In both cases, the Young's modulus increased gradually in the first 1 h and then became smooth in the next 1 h. We can see that the Young's modulus of B-lymphoma cells without rituximab was approximately five times larger than that stimulated by rituximab. This indicated that cells became significantly softer after incubation with rituximab.

Many researchers have used nanorobots to monitor the changes in mechanical properties on incubation of the cells with drugs [68–70]. From Fig. 8.15, we can see that B-lymphoma cells became markedly softer after being incubated with rituximab. Rituximab can specifically bind to the CD20 on the B-lymphoma cell surface. The binding to the CD20 can induces a rapid translocation of CD20 molecules to lipid rafts [6], as shown in Fig. 8.16. Lipid rafts are fluctuating nanoscale assemblies of sphingolipid, cholesterol, and proteins that can be stabilized to coalesce, forming platforms that function in membrane signaling and trafficking [78]. Then the interaction between the redistributed CD20 molecules and the raft membrane protein components results in the activation of the transmembrane signaling machinery [79]. Unruh et al. [80] have demonstrated that the translocation and hypercrosslinking of CD20 are necessary for rituximab to

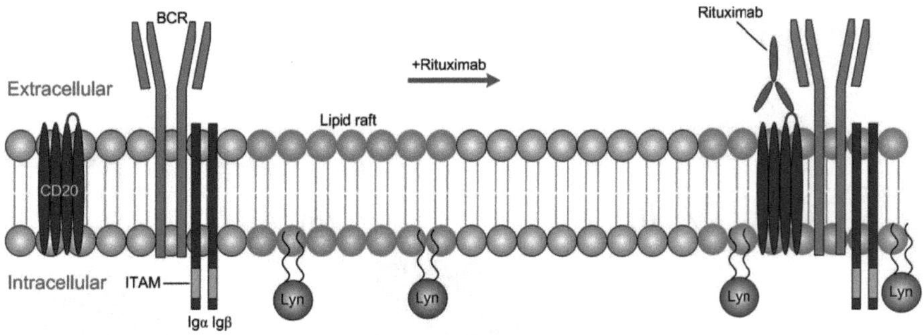

Fig. 8.16 Schematic diagram of rituximab-mediated changes of cell membrane [62]. Rituximab binding to CD20 induces the translocation of CD20 to lipid rafts and results in the tight association of CD20 and BCR

activate src family kinases (SFK)-dependent signaling and apoptosis. Janas et al. [81] have shown that rituximab-induced translocation of CD20 to lipid rafts is crucial for increased intracellular Ca^{2+} levels and downstream apoptotic signaling. Also, the rituximab binding to CD20 significantly interferes with B-cell receptor (BCR) signaling, resulting in a time-dependent inhibition of the BCR-signaling cascade involving Lyn, Syk, PLCc2, Akt, and ERK, and calcium mobilization [82]. Walsh et al. [83] have suggested that after the ligation of CD20 by rituximab, CD20 molecules cluster together and associate with the BCR, as shown in Fig. 8.16. In other words, the binding of rituximab to CD20 results in changes in the cell membrane. Hence the changes in the cell membrane may relate to the decrease of the Young's modulus. On the other hand, the binding of rituximab to CD20 can activate the apoptotic signaling pathways. As the apoptosis pathway promotes, some changes on the cell take place and these changes may relate to the reduction of the Young's modulus. Hu et al. [71] have shown that the Young's modulus of apoptotic lymphocytes was dramatically smaller than that of activated lymphocytes. Cai et al. [72] have shown that the apoptotic human pterygium fibroblasts (HPF) appeared 70 % softer compared to the normal HPFs. Finally, the cellular mechanical properties are dependent on the cytoskeleton [84]. Any changes in the cytoskeleton will result in changes in the mechanical properties. The binding of rituximab to CD20 may induce the changes in the cytoskeleton which lead to the softening of the mechanical properties. Altogether, so far there is no available information concerning the relationship between the mechanical properties of B-lymphoma cells and the rituximab's in vivo mechanisms, and further experiments are needed for elucidating the underlying mechanisms.

For the future work, we would like to separate living lymphoma patient tumor cells from normal cells and then investigate their mechanical properties in order to build the relationship between the mechanical properties of lymphoma cells and rituximab's efficacies, potentially providing a rapid and high-efficiency detection method for predicting the efficacy of the targeted therapy.

8.5 Measurement of Protein–Protein Binding Forces and Protein Distribution

Using the functionalized tip to obtain force curves on the protein-coated substrate, the binding forces of antibody-antigen/receptor-ligand can be measured. SMFS was first performed in the mid-1990s. In 1994, Florin et al. [43] measured the binding force of biotin–avidin; in 1996, Hinterdoefer et al. [46] measured the binding forces of human serum albumin (HSA) and anti-HSA, by linking anti-HSA onto nanorobot tip with PEG linker molecules and linking HSA onto mica surface. Besides measuring the molecular forces on protein-coated substrate, SMFS can also measure the molecular forces on cell surface. In 2006, Puntheeranurak et al. [54] measured the specific binding forces of Na + -glucose co-transporter(SGLT1) and its antibodies on Chinese hamster ovary(CHO) cells; in 2009, Shi et al. [85] investigated the interaction forces of Heregulin and HER3 on human embryonic kidney (HEK) 293 cells.

The mechanism of rituximab's different efficacies is still unknown. We measured the binding force of CD20 and rituximab to provide a new exploration for the different efficacies. The experiment was first performed on purified CD20 proteins, and then on lymphoma Raji cells. Rituximab molecules were linked onto the nanorobot tip through NHS-PEG-MAL linker molecule according to the procedure presented in Sect. 8.2.3.

SEM and fluorescence microscopy (FM) were applied to verify that rituximab had been linked onto nanorobot tips. For SEM, the functionalized probe was attached onto a copper puck. Then the puck was place onto the stage of the SEM (Nova NanoSEM, Institute of Metal Research Chinese Academy of Sciences) for being scanned. The high voltage was 20 kV. The magnification was 30,000 × for large-scale scanning and 80,000 × for small-scale scanning. For control experiments, the unfunctionalized probe was scanned the same as the functionalized probe. For FM, the RBITC-labeled goat anti-human IgG (Solarbio company, China) was used. Rituximab binds CD20 on the nanorobot probe and RBITC-labeled goat anti-human IgG binds rituximab. First the functionalized probe was placed into a petri dish in PBS and then rituximab solution was added and incubated for 1 h. The unbound rituximabs were removed by PBS. Then the RBITC-labeled goat anti-human IgG solution was added and incubated for 1 h. After the reaction the functionalized probe was rinsed with PBS to remove the unbound IgG and then placed onto the stage of the fluorescence microscope. The green excitation light was used. For control experiments, the unfunctionalized probe was treated the same as the functionalized probe.

Figure 8.17 shows the scanning electron microscopy (SEM) and FM images of verification experiments that rituximabs had tethered onto the nanorobot tips. Figure 8.17a is the SEM image of an unfunctionalized tip and Fig. 8.17b is the higher resolution SEM image by zooming into the local area of the tip. Figure 8.17c is the SEM image of a functionalized tip and Fig. 8.17d is the higher resolution SEM image. We can see that there were many particles on the surface of

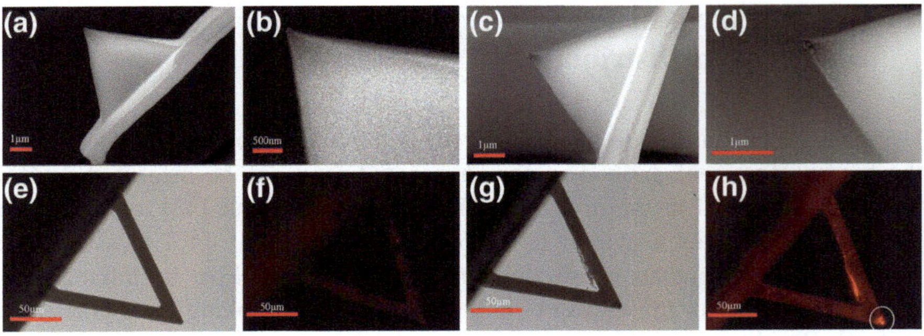

Fig. 8.17 Contrast of unfunctionalized probes and functionalized probes [86]. **a**, **b** SEM images of the unfunctionalized probe. **c**, **d** SEM images of the functionalized probe. **e** Optical image of the unfunctionalized probe. **f** Secondary antibody fluorescence labeling image of the unfunctionalized probe. **g** Optical image of the functionalized probe. **h** Secondary antibody fluorescence labeling image of the functionalized probe

the functionalized tip while the surface of the unfunctionalized tip was smooth. The optical images of an unfunctionalized tip and a functionalized tip are shown in Fig. 8.17e and g, respectively. The secondary antibody labeling fluorescence images of the unfunctionalized tip and the functionalized tip are shown in Fig. 8.17f and h, respectively. We can see that the fluorescence around the tip (denoted by white circle in Fig. 8.17h) is obviously brighter than other areas, while the fluorescence of the unfunctionalized tip is dim. The nanorobot tip used here was pyramidal and the height of the tip was in the range of 2.5–8.0 μm. After the functionalized procedure, rituximabs bound onto the tip surface through PEG linker, as seen in Fig. 8.17d, and the tip surface was covered by a large number of particles. Rituximabs should also bind onto the cantilever of the probe. But the density of rituximabs on the tip was much larger than in the cantilever. Then after the secondary antibody labeling fluorescence experiments, the number of secondary antibody on the tip was much more than on the cantilever. This resulted in the fluorescence of the tip being markedly brighter than that of the cantilever.

CD20 proteins were attached onto freshly cleaved mica (1 cm × 1 cm squares) according to the established procedure [87]. ① Purge a glass desiccator with argon for 2 min. ② Pipette 30 μL APTES into a small petri dish and 10 μL NN-diisopropylethylamine into another small petri dish. ③ After putting the freshly cleaved mica and the two petri dishes into the glass desiccator, purge the desiccator with argon for 2 min and seal for 0.5–2 h. ④ Pipette 200 μL glutaraldehyde solution onto the mica and incubated for 10 min. ⑤ Rinse the mica with Milli-Q ultra-pure water and then add 100 μL CD20 protein solution onto the mica and incubate for 30 min. ⑥ Rinse the mica with Milli-Q ultra-pure water again and put the mica into a petri dish with PBS for being probed by a nanorobot.

Figure 8.18a, b shows the nanorobot contact mode images of CD20 proteins. Figure 8.18a is the topography image, and Fig. 8.18b is the deflection image. We can see many CD20 protein particles distributed on the mica surface. The prerequisite of imaging proteins in fluids is attaching them onto a flat support [88].

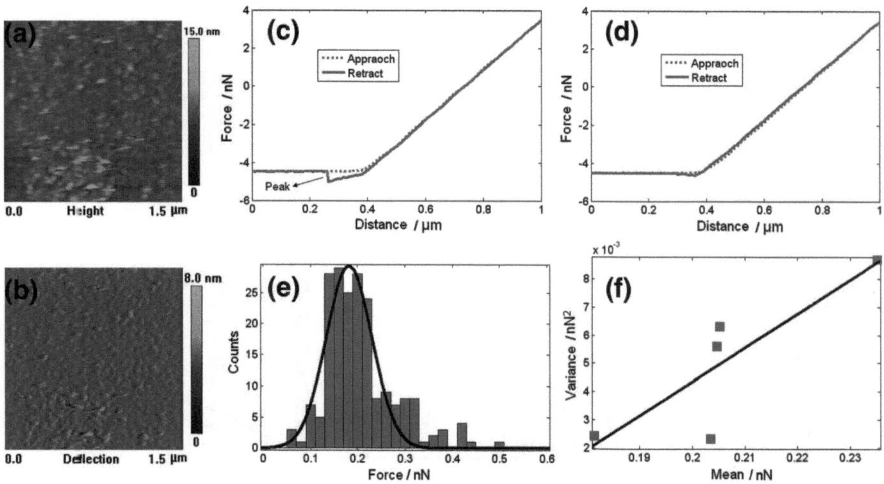

Fig. 8.18 Measuring the rupture force between purified CD20 proteins and rituximab [94]. **a** Nanorobot contact mode topography image of CD20 proteins. **b** Nanorobot contact mode deflection image of CD20 proteins. **c** A typical force curve. **d** A force curve after adding free rituximab. **e** Histogram of the rupture forces. **f** Poisson analysis was used to compute the individual rupture force

Owing to the atomic flat surfaces and easy preparation, mica was the most commonly used support [89]. The intrinsic hydrophobicity of proteins results in that they are difficult to attach to the hydrophilic mica [90]. Various methods, including physical adsorption and covalent attachment, have been provided for attaching proteins onto mica [91]. Henderson et al. [90] have imaged single ROMK1 proteins by coating the mica with cetylpyridinium chloride. Kada et al. [92] have used nanorobots to visualize single ryanodine receptors which were adsorbed onto mica surfaces in the presence of Ca^{2+}. Wang et al. [93] have imaged IgG_4 proteins by treating the mica with APTES and glutaraldehyde. However, because of the soft, fragile, corrugated, and dynamic nature of the cell surface, visualizing the topography of single receptors on living mammalian cell surface is still a big challenge [21] and teamwork between the experts from different scientific disciplines is essential for the goal.

Figure 8.18c–f represents the results of measuring the binding forces between purified CD20 proteins and rituximab. A typical force curve obtained on CD20-coated mica with rituximab-conjugated probes is shown in Fig. 8.18c. The black line indicates the approach process, and the red line indicates the retract process. We can see that there was an abrupt peak on the retract curve. When performing the functionalization experiments, the bonds between nanorobot tip and PEG linker molecule, the bonds between rituximab and PEG linker molecule, and the bonds between CD20 and substrate should be much stronger than the bonds between CD20 and rituximab. Researches indicate that the bonds between the NHS end of the PEG linker and nanorobot tip were much stronger than that between protein–protein [42, 46]. Additionally, the bonds between the MAL end

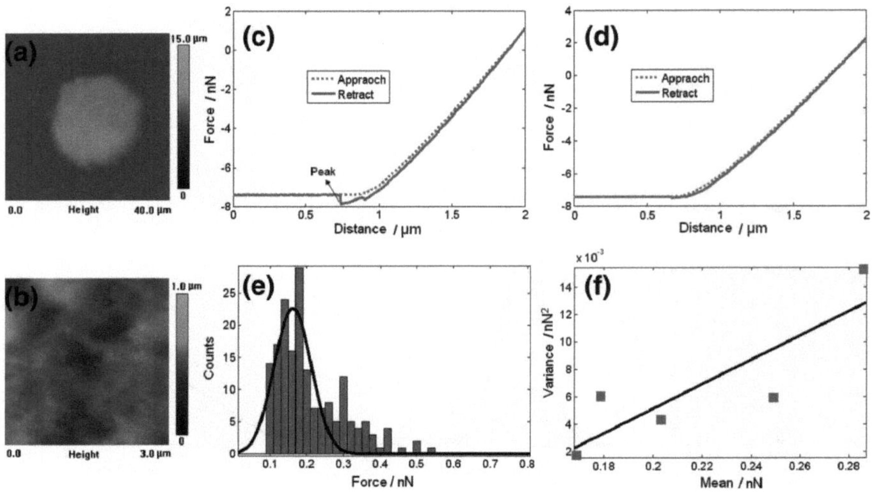

Fig. 8.19 Measuring the rupture force between CD20 proteins on the lymphoma Raji cells and rituximab [94]. **a** Nanorobot contact mode topography image of Raji cell. **b** Nanorobot contact mode topography image of the local area of the Raji cell. **c** A typical *force curve*. **d** A *force curve* after adding free rituximab. **e** Histogram of the rupture forces. **f** Poisson analysis was used to compute the individual rupture force

of the PEG linker and antibody and the bonds between proteins and aldehyde group on the functionalized mica were much stronger than that between protein and proteins [48, 93]. Therefore, during the process of stretching the CD20-rituximab complex at force curve mode, CD20-rituximab complex first ruptures. After adding free rituximabs, force curves were obtained again and there were no peaks on the retract curve, as shown in Fig. 8.18d. This verifies that the peak in Fig. 8.18c corresponded to the specific binding between CD20 proteins on the mica and the rituximab on the nanorobot tip. After computing the rupture forces for each force curve obtained on the mica, the histogram of the rupture forces was constructed (Fig. 8.18e). Gaussian fit of the histogram indicated that the rupture force was 182 $\pm$ 69 pN. For the four measurements, mean and variance were calculated. The mean was 0.0063, 0.0024, 0.0023, 0.0087, respectively, and the variance was 0.0063, 0.0024, 0.0023, 0.0087, respectively. The fitting function is $f(x) = p_1{*}x + p_2$ and the fitting results indicate that $p_1 = 0.1213, p_2 = -0.02011$. We can see that the slope of the fitting line indicated that the individual rupture force was 121pN (Fig. 8.18f).

The Burkitt's lymphoma Raji cells were cultured at 37 °C (5 %CO_2) in RPMI-1640 culture medium containing 10 % fetal bovine serum. Raji cells were attached on the glass slide by poly-L-lysine's electrostatic adsorption and glutaraldehyde's chemical fixation. Then the functionalized probes were used to measure the CD20-rituximab binding forces on the fixed Raji cells. Figure 8.19 shows the results of measuring the rupture forces between CD20 proteins on the lymphoma Raji cells and rituximab. Figure 8.19a was the contact mode topography image. The scan size was 40 µm. To visualize the topography of the cell surface's local area, scan

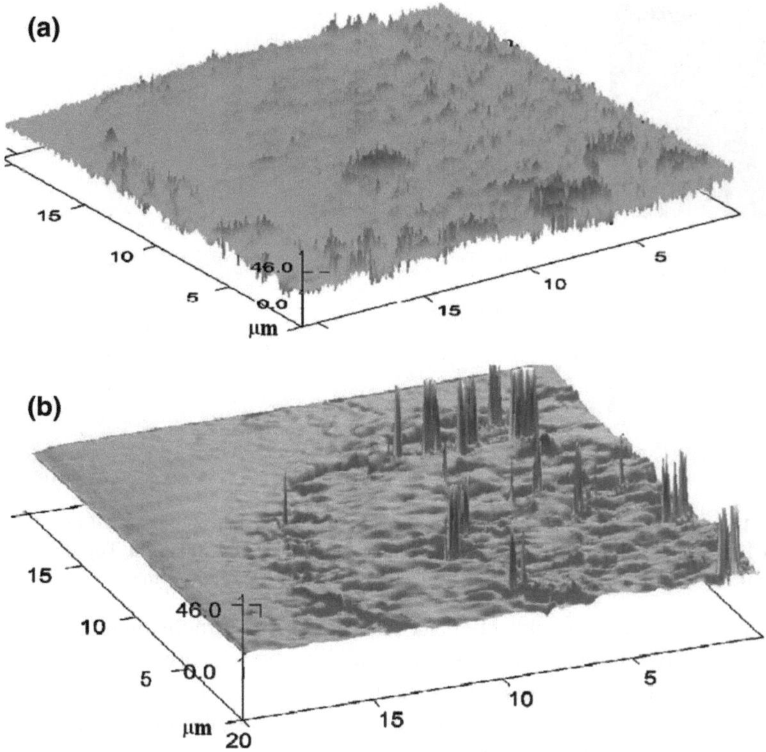

Fig. 8.20 Visualizing the distribution of CD20 on the cell surface with lift mode method [61]. Three-dimensional phase image with the normal tip (**a**) and with the functionalized tip (**b**)

size was reduced to 3 μm and Fig. 8.19b was the topography image of the local area. From the images of the local area, we can see that the cell surface was corrugated and rough. A typical force curve obtained on lymphoma Raji cells with rituximab-conjugated probes is shown in Fig. 8.19c. We can see that there was an abrupt peak on the retract curve, and after adding free rituximab, the peak disappeared (Fig. 8.19d). This verified the CD20-rituximab-specific binding interactions. Figure 8.19e is the histogram of the rupture forces. Gaussian fit indicated that the rupture force was 162 ± 71pN. For the five measurements, the mean was 0.1787, 0.2859, 0.2033, 0.1690, 0.2493, respectively, and the variance was 0.0060, 0.0153, 0.0043, 0.0017, 0.0059, respectively. The fitting function is $f(x) = p_1{*}x + p_2$ and the fitting results indicate that $p_1 = 0.08888$, $p_2 = -0.01267$. The slope of the fitting line of the plot of variance against mean indicated that the individual rupture force was 89pN (Fig. 8.19f).

The distribution of CD20 on the lymphoma cell surface was visualized with the functionalized tip in the lift mode (Fig. 8.20). When the normal tip was used to scan the surface of lymphoma cells to obtain the phase image in the lift mode,

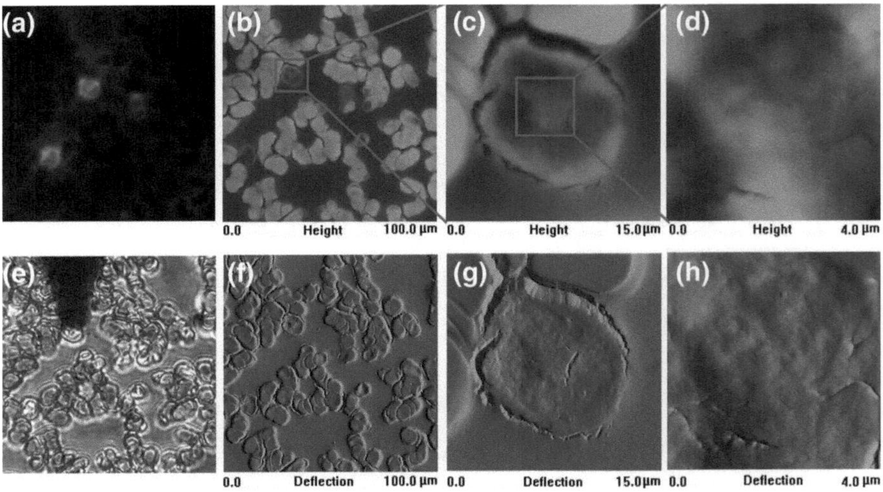

Fig. 8.21 Imaging the topography of the cancer cells using nanorobots guided by fluorescence [86]. **a** Fluorescence image. **b** Topography image. **c** Topography image. **d** Topography image. **e** Optical image. **f** Deflection image. **g** Deflection image. **h** Deflection image

there was no distinct abrupt peak in the three-dimensional phase image (Fig. 8.20a). By contrast, when the functionalized tip was used to obtain the phase image, there were many distinct abrupt peaks in the three-dimensional phase image (Fig. 8.20b). These peaks correspond to the distribution of CD20.

In summary, with the probe functionalization technology, we can quantitatively measure the binding forces between individual protein and proteins, and visualize the distribution of proteins on the cell surface. This can provide novel information about understanding the rituximab's mechanisms from the perspective of biomechanics.

8.6 Measuring the CD20-Rituximab Binding Force on Lymphoma Patient Cells

The patients' cells were prepared from the bone marrow of five B-cell lymphoma patients whose bone marrow was intruded by lymphoma cells. Poly-L-lysine was used to immobilize cells onto a glass slide. The poly-L-lysine (0.1 %) was diluted with Milli-Q ultra-pure water to one-tenth of the original value and then was dropped onto fresh glass slides. These slides were stored at room temperature overnight. The paracentesis needle was used to acquire the bone marrow from the lymphoma patients. Then the bone marrow was dropped onto the poly-L-lysine-coated glass slides. A fresh glass slide was used to disperse the bone marrow and 4 % formaldehyde was used for chemical fixation.

For cancer cell recognition, rituximab and RBITC-labeled goat anti-human IgG were used. ① Put the slide sample of lymphoma patients into a petri dish with 10 mL PBS. Add 1 mL rituximab into the petri dish and incubate in 4 °C for 4 h.

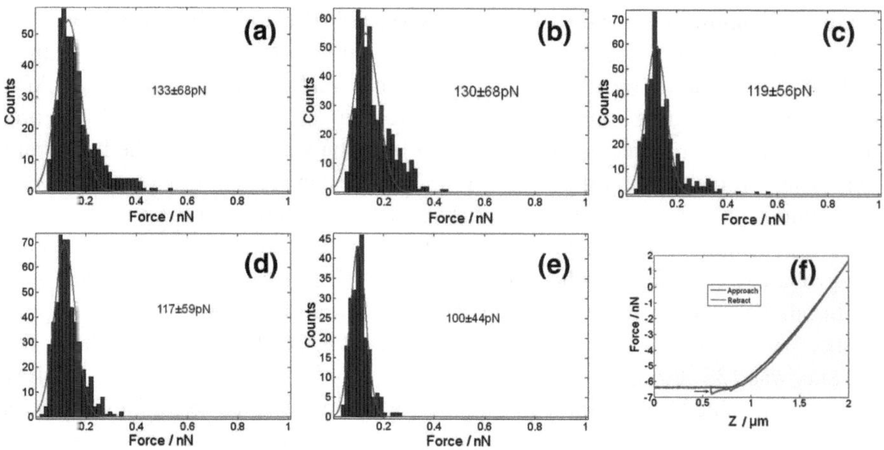

Fig. 8.22 Measuring the CD20 rituximab binding forces on patients' cells from five lymphoma patients. **a** Histogram of binding forces of patient 1. **b** Histogram of binding forces of patient 2. **c** Histogram of binding forces of patient 3. **d** Histogram of binding forces of patient 4. **e** Histogram of binding forces of patient 5. **f** Typical *force curves*

② Wash the sample thrice with PBS.③ Drop the solution of RBITC-labeled goat anti-human IgG onto the sample and incubate in 4 °C for 1 h. ④ Wash the sample thrice with PBS. ⑤ Put the sample onto the stage of the fluorescence microscope and use the green excitation light.

Guided by the fluorescence, the detailed topography of cancer cells were imaged by nanorobots, as shown in Fig. 8.21. From the fluorescence image (Fig. 8 21a), we can see a cancer cell (denoted by the circle). Then the probe was moved to the cancer cell. From the optical image (Fig. 8.21e), we can see that the probe had been moved to the cancer cell. First, a large scan (100 μm) was performed, and the topography image (Fig. 8.21b) and deflection image (Fig. 8.21f) were obtained. The cancer cell can easily be recognized in the nanorobot images (denoted by the square in Fig. 8.21b), and higher resolution topography image (Fig. 8 21c) and deflection image (Fig. 8.21g) of the cancer cell were obtained. By zooming into the local area (4 μm) of the cell surface, the ultrastructure of the cell surface was observed (Fig. 8.21d, h). From the nanorobot images of the cancer cell, we can see that cancer cells exhibit different morphology, compared with the other cells on the sample. The other cells have smooth and flat surfaces, while the cancer cells have relatively corrugated and hollow surfaces. The diameter of the cancer cell measured with nanorobot offline software (Version 6.13) is about 8.78 μm, while the typical diameter of the other cells is about 5.73 μm, as shown in Fig. 8.21f. We can see that the diameter of the cancer cell is clearly larger than that of the other cells.

In order to measure the CD20-rituximab binding forces on pathologic cell sample, the method used here was using functionalized probes to obtain force curves on cells from the sample. If the force curves had prominent abrupt peaks, then the cell was chosen for measuring the CD20-rituximab binding forces.

Figure 8.22 shows the results of measuring the CD20-rituximab binding forces on the patients' cells from five lymphoma patients. For each patient's sample, ~1,000 force curves were obtained at the same loading rate of 1.99 μm/s. After computing the binding forces from these force curves, the histogram of the force distribution was constructed. The Gaussian function was used to fit the histogram. One typical force curve is shown in Fig. 8.22f. We can see that there was one well-pronounced abrupt peak in the force curve. During the retract process of the probe, the CD20-rituximab complexes were stretched by the probe-exerted force. When the probe-exerted force reached the strength of CD20-rituximab bonds, the bonds ruptured. The abrupt peak indicated the CD20-rituximab complex bond rupture events. According to the Gaussian fit results, the binding force of these five patients was 133 ± 68pN, 130 ± 68pN, 119 ± 56pN, 117 ± 59pN, 100 ± 44pN, respectively. The binding force of patient 1 was almost equal to the binding force of patient 2. The binding force of patient 2 was 11pN larger than the binding force of patient 3. The binding force of patient 3 was almost equal to the binding force of patient 4 and the binding force of patient 4 was 17pN larger than the binding force of patient 5.

As shown in Fig. 8.22, we can see that the CD20-rituximab binding forces were a little variable among different lymphoma patients. The information of the five lymphoma patients is shown in Table 8.1. There were two men of ages 60 and 52 years respectively. The ages of the other three women were 70, 69, and 62 years, respectively. They were all B-cell lymphomas. As with B-cell lymphoma, the optimal results are received when rituximab is combined with chemotherapy [95]. The clinical treatments of the five patients were mainly combination therapy (chemotherapy + rituximab). Besides chemotherapy, patients 1, 2, and 4 were simultaneously treated with rituximab for 6 cycles, 1 cycle, and 6 cycles, respectively. Patients 3 and 5 were only treated with chemotherapy. We can see that the five binding forces can be divided into three categories according to the subtype of the lymphoma. Patients 1 and 2 belong to the first category; patients 3 and 5 belong to the second category; patient 4 belongs to the last category. We can see that the binding force of patient 1 was equal to the binding force of patient 2 and both these patients were small B cell non-Hodgkin's lymphoma/chronic lymphocytic leukemia. Patients 3 and 4 also had the same subtype (follicular B-cell lymphoma), but they had different binding forces. This may be due to the individual differences in patients.

The patient responses of the five lymphoma patients are also shown in Table 8.1. The patient responses of patients 2 and 5 were not evaluated owing to the fewer treatment cycles. We can see that patient 3 had partial remission after the treatment of chemotherapy, while patient 4 had complete remission after the combination treatment of chemotherapy and 6 cycles' rituximab therapy. This indicates that the combination therapy can achieve better therapeutic effects than monotherapy. Patient 1 was also treated with 6 cycles' rituximab therapy, but the response was stable. Comparing patient 1 with patient 4, we can see that the same treatment for different lymphoma patients results in variable patient responses. The CD20-rituximab binding force of patient 1 was larger than that of patient 4, but the efficacy of

Table 8.1 Patient characteristics and clinical therapy versus force measurements

Case no.	Age	Sex	Tumorsubtype	Rituximab therapy	Chemotherapy	Patient response	Force (PN)
1	60	Male	Small B cell non-Hodgkin's lymphoma/chronic lymphocytic leukemia	6 cycles	+	Stable	133 ± 68
2	70	Female	Small B cell non-Hodgkin's lymphoma/chronic lymphocytic leukemia	1 cycle		Unevaluated	130 ± 68
3	69	Female	Follicular non-Hodgkin's lymphoma	–	+	Partial remission	119 ± 56
4	62	Female	Splenic marginal zone B-cell lymphoma/chronic lymphocytic leukemia	6 cycles	+	Complete remission	117 ± 59
5	52	Male	Follicular B-cell lymphoma	+	–	Unevaluated	100 ± 44

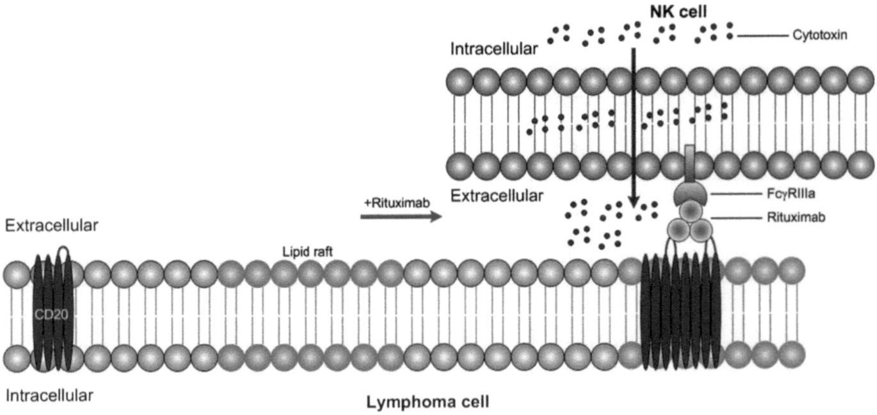

Fig. 8.23 ADCC-mediated mechanism of killing the lymphoma cells

patient 1 was worse than that of patient 4. This may be due to the rituximab's killing mechanisms. Rituximab is a Type 1 anti-CD20 mAb, of which the main mechanisms to kill the targeted cells were CDC and ADCC [96]. It is widely accepted that ADCC is critical for the success of anti-CD20 immunotherapy, while CDC is still disputed, and in vitro experiments indicate that natural killer (NK) cell is prominent in ADCC [7]. NK cells express the FcγRIIIa which can bind antibody's Fc domains [4]. As illustrated in Fig. 8.23, after the CD20-rituximab binding, the CD20 protein translocates to the lipid rafts [6] and the Fc domain of rituximab binds to the FcγRIIIa of NK cells, which triggers the release of cytotoxin to lyse the tumor cell [4]. Some patients can express a low affinity version of the Fc receptor on their immune effector cells [77]. Hence the affinity of Fc receptors of patient 1 may be lower than that of patient 4 and this difference results in the different rituximab's efficacies. Altogether, rituximab's in vivo mechanism is still unclear and it is hard to interpret the individual differences of patients and further experiments are strongly needed for unraveling the underlying mechanisms.

8.7 Summary

The clinical experiences of rituximab has provided a wealth of data about therapeutic efficacies since its approval in 1997 and boosted the advancement of mAb. However, the mAb-based therapy is still in its infancy and more work is urgently needed to improve its efficacy. Also, biomarkers that can be used for predicting the targeted drug efficacy or for early cancer detection are strongly needed. The invention of nanorobotics provides extraordinary opportunities to meet the challenges of targeted therapy. The rapid developments in nanorobots have allowed researchers to characterize the behavior and measure the interactions of single cells and single molecules, providing new insights into the mechanisms of cellular processes compared to traditional ensemble experiments and bringing new techniques for discovering novel biomarkers. Using nanorobotics techniques to fill the gap between cellular activities at individual cell/molecule levels and clinical practice will have a great impact on disease treatment and drug development, potentially developing a rapid and high-efficiency detection method for molecular targeted therapy.

Acknowledgments This work was supported by the National Natural Science Foundation of China (Project No.60904095, 61175103), and the CAS FEA International Partnership Program for Creative Research Teams.

References

1. Jemal A, Siegel R, Xu J, Ward E (2010) Cancer statistics, 2010. CA Cancer J Clin 60:277–300
2. Schrama D, Reisfeld RA, Becker JC (2006) Antibody targeted drugs as cancer therapeutics. Nat Rev Drug Discov 5:147–159
3. Sawyers C (2004) Targeted cancer therapy. Nature 432:294–297
4. Adams GP, Weiner LM (2005) Monoclonal antibody therapy of cancer. Nat Biotechnol 23:1147–1157
5. Cheson BD, Leonard JP (2008) Monoclonal antibody therapy for B-cell non-Hodgkin's lymphoma. N Eng J Med 359:613–626
6. Cartron G, Watier H, Golay J, Solal-Celigny P (2004) From the bench to the bedside: ways to improve rituximab efficacy. Blood 104:2635–2642
7. Lim SH, Beers SA, French RR, Johnson PWM, Glennie MJ, Cragg MS (2010) Anti-CD20 monoclonal antibodies: historical and future perspectives. Haematologica 95:135–143
8. Nimmerjahn F, Ravetch JV (2007) Antibodies, Fc receptors and cancer. Curr Opin Immunol 19:239–245
9. Chan AC, Carter PJ (2010) Therapeutic antibodies for autoimmunity and inflammation. Nat Rev Immunol 10:301–316
10. Carter P (2001) Improving the efficacy of antibody-based cancer therapies. Nat Rev Cancer 1:118–129
11. Varmus H (2006) The new era in cancer research. Science 312:1162–1165
12. Beers SA et al (2010) Antigenic modulation limits the efficacy of anti-CD20 antibodies: implications for antibody selection. Blood 115:5191–5201
13. Sawyers CL (2008) The cancer biomarker problem. Nature 452:548–552
14. Srinivas PR, Kramer BS, Srivastava S (2001) Trends in biomarker research for cancer detection. Lancet Oncol 2:698–704
15. Etzioni R, Urban N, Ramsey S, McIntosh M, Schwartz S, Reid B, Radich J, Anderson G, Hartwell L (2003) The case for early detection. Nat Rev Cancer 3:1–10
16. Zhuang X, Bartley LE, Babcock HP, Russell R, Ha T, Herschlag D, Chu S (2000) A single-molecule study of RNA catalysis and folding. Science 288:2048–2051
17. Cecconi C, Shank EA, Bustamante C, Marqusee S (2005) Direct observation of the three-state folding of a single protein molecule. Science 309:2057–2060
18. Xie XS, Yu J, Yang WY (2006) Living cells as test tubes. Science 312:228–230
19. Dufrene YF (2009) Atomic force microscopy: a powerful molecular toolkit in nanoproteomics. Proteomics 9:5400–5405
20. Dupres V, Alsteens D, Andre G, Verbelen C, Dufrene YF (2009) Fishing single molecules on live cells. Nano Today 4:262–268
21. Muller DJ, Dufrene YF (2011) Force nanoscopy of living cells. Curr Biol 21:R212–R216
22. Dufrene YF, Evans E, Engel A, Helenius J, Gaub HE, Muller DJ (2011) Five challenges to bringing single-molecule force spectroscopy into living cells. Nat Methods 8:123–127

23. Armani AM, Kulkarni RP, Fraser SE, Flagan RC, Vahala KJ (2007) Label-free, single-molecule detection with optical microcavities. Science 317:783–787
24. Giepmans BNG, Adams SR, Ellisman MH, Tsien RY (2006) The fluorescent toolbox for assessing protein location and function. Science 312:217–224
25. Yu X, Xu D, Cheng Q (2006) Label-free detection methods for protein microarrays. Proteomics 6:5493–5503
26. Ona T, Shibata J (2010) Advanced dynamic monitoring of cellular status using label-free and non-invasive cell-based sensing technology for the prediction of anticancer drug efficacy. Anal Bioanal Chem 398:2505–2533
27. Neuman KC, Nagy A (2008) Single-molecule force spectroscopy: optical tweezers, magnetic tweezers and atomic force microscopy. Nat Methods 5:491–505
28. Walter NG, Huang CY, Manzo AJ, Sobhy MA (2008) Do-it-yourself guide: how to use the modern single-molecule toolkit. Nat Methods 5:475–489
29. Muller DJ, Dufrene YF (2008) Atomic force microscopy as a multifunctional molecular toolbox in nanobiotechnology. Nat Nanotechnol 3:261–269
30. Sitti M (2004) Micro-and nano-scale robotics. In: Proceedings of American Control Conference, pp 1–8
31. Xi N, Fung CKM, Yang R, Seiffert-Sinha K, Lai KWC, Sinha AA (2010) Nanomanipulation using atomic force microscopy. IEEE Nanotechnol Mag 4(1):9–12
32. Dong L, Nelson BJ (2007) Robotics in the small part II: nanorobotics. IEEE Robot Autom Mag 14:111–121
33. Patel GM, Patel GC, Patel RB, Patel JK, Patel M (2006) Nanorobot: a versatile tool in nanomedicine. J Drug Target 14:63–67
34. Freitas RA (2005) What is nanomedicine? Nanomedicine NBM 1:2–9
35. Cavalcanti A, Shirinzadeh B, Kretly LC (2008) Medical nanorobotics for diabetes control. Nanomedicine NBM 4:127–138
36. Hill C, Amodeo A, Joseph JV, Patel HRH (2008) Nano-and microrobotics: how far is the reality. Expert Rev Anticancer Ther 8(12):1891–1897
37. Binnig G, Quate CF, Gerber C (1986) Atomic force microscope. Phys Rev Lett 56:930–933
38. Li G, Xi N, Yu M, Fung WK (2004) Development of augmented reality system for AFM-based nanomanipulation. IEEE-ASME Trans Mechatron 9:358–365
39. Sitti M (2001) Survey of nanomanipulation systems. In: Proceedings of IEEE International Conference on Nanotechnology, pp 75–80
40. Sitti M (2003) Teleoperated and automatic nanomanipulation systems using atomic force microscope probes. Proc IEEE Conf Decis, Control, pp 2118–2123
41. Touhami A, Nysten B, Dufrene YF (2003) Nanoscale mapping of the elasticity of microbial cells by atomic force microscopy. Langmuir 19:4539–4543
42. Hinterdorfer P, Dufrene YF (2006) Detection and localization of single molecular recognition events using atomic force microscopy. Nat Methods 3:347–355
43. Florin EL, Moy VT, Gaub HE (1994) Adhesion forces between individual ligand-receptor pairs. Science 1994:415–417
44. Fritz J, Katopodis AG, Kolbinger F, Anselmetti D (1998) Force-mediated kinetics of single P-selectin/ligand complexes observed by atomic force microscopy. Proc Natl Acad Sci USA 95:12283–12288
45. Dupres V et al (2005) Nanoscale mapping and functional analysis of individual adhesions on living bacteria. Nat Methods 2:515–520
46. Hinterdorfer P, Baumgartner W, Gruber HJ, Schilcher K, Schindler H (1996) Detection and localization of individual antibody-antigen recognition events by atomic force microscopy. Proc Natl Acad Sci USA 93:3477–3481
47. Stroh C, Wang H, Bash R, Ashcroft B, Nelson J, Gruber H, Lohr D, Lindsay SM, Hinterdorfer P (2004) Single-molecule recognition imaging microscopy. Proc Natl Acad Sci USA 101:12503–12507
48. Ebner A et al (2007) A new, simple method for linking of antibodies to atomic force microscopy tips. Bioconjugate Chem 18:1176–1184

49. Muller DJ, Helenius J, Alsteens D, Dufrene YF (2009) Force probing surfaces of living cells to molecular resolution. Nat Chem Biol 5:383–390
50. Li G, Xi N, Wang DH (2005) Investigation of angiotensin II type 1 receptor by atomic force microscopy with functionalized tip. Nanomedicine NBM 1:306–312
51. Butt HJ, Wolff EK, Gould SAC, Northern BD, Peterson CM, Hansma PK (1990) Imaging cells with the atomic force microscope. J Struct Biol 105:54–61
52. Deng Z, Lulevich V, Liu F, Liu G (2010) Applications of atomic force microscopy in biophysical chemistry of cells. J Phys Chem B 114:5971–5982
53. Matzke R, Jacobson K, Radmacher M (2001) Direct, high-resolution measurement of furrow stiffening during division of adherent cells. Nat Cell Biol 3:607–610
54. Puntheeranurak T, Wildling L, Gruber HJ, Kinne RKH, Hinterdorfer P (2006) Ligands on the string:single-molecule AFM studies on the interaction of antibodies and substrates with the Na^+-glucose co-transporter SGLT1 in living cells. J Cell Sci 119:2960–2967
55. Fantner GE, Barbero RJ, Gray DS, Belcher AM (2010) Kinetics of antimicrobial peptide activity measured on individual bacterial cells using high-speed atomic force microscopy. Nat Nanotechnol 5:280–285
56. Kasas S, Ikai A (1995) A method for anchoring round shaped cells for atomic force microscope imaging. Biophys J 68:1678–1680
57. Dufrene YF (2008) Atomic force microscopy and chemical force microscopy of microbial cells. Nat Protoc 3:1132–1138
58. Rosenbluth MJ, Lam WA, Fletcher DA (2006) Force microscopy of nonadherent cells: a comparison of leukemia cell deformability. Biophys J 90:2994–3003
59. Ng L, Hung HH, Sprunt A, Chubinskaya S, Ortiz C, Grodzinsky A (2007) Nanomechanical properties of individual chondrocytes and their developing growth factor-stimulated pericellular matrix. J Biomech 40:1011–1023
60. Jena BP, Cho SJ (2002) The atomic force microscope in the study of membrane fusion and exocytosis. Method Cell Biol 68:33–50
61. Li M, Liu L, Xi N, Wang Y, Dong Z, Li G, Xiao X, Zhang W (2010) Detecting CD20-Rituximab specific interactions on lymphoma cells using atomic force microscopy. Sci China Life Sci 53:1189–1195
62. Li M, Liu L, Xi N, Wang Y, Dong Z, Tabata O, Xiao X, Zhang W (2011) Imaging and measuring the rituximab-induced changes of mechanical properties in B-lymphoma cells using atomic force microscopy. Biochem Biophys Res Commun 404:689–694
63. Lee GYH, Lim CT (2007) Biomechanics approaches to studying human diseases. Trends Biotechnol 25:111–118
64. Bao G, Suresh S (2003) Cell and molecular mechanics of biological materials. Nat Mater 2:715–725
65. Cross SE, Jin YS, Rao J, Gimzewski JK (2007) Nanomechanical analysis of cells from cancer patients. Nat Nanotechnol 2:780–783
66. Li QS, Lee GYH, Ong CN, Lim CT (2008) AFM indentation study of breast cancer cells. Biochem Biophys Res Commun 374:609–613
67. Rotsch C, Jacobson K, Radmacher M (1999) Dimensional and mechanical dynamics of active and stable edges in motile fibroblasts investigated by using atomic force microscopy. Proc Natl Acad Sci USA 96:921–926
68. Rotsch C, Radmacher M (2000) Drug-induced changes of cytoskeletal structure and mechanics in fibroblasts: an atomic force microscopy study. Biophys J 78:520–535
69. Lam WA, Rosenbluth MJ, Fletcher DA (2007) Chemotherapy exposure increases leukemia cell stiffness. Blood 109:3505–3508
70. Martens JC, Radmacher M (2008) Softening of the actin cytoskeleton by inhibition of myosin II. Pflugers Arch Eur J Physiol 456:95–100
71. Hu M, Wang J, Zhao H, Dong S, Cai J (2009) Nanostructure and nanomechanics analysis of lymphocyte using AFM: from resting, activated to apoptosis. J Biomech 42:1513–1519
72. Cai X, Yang X, Cai J, Wu S, Chen Q (2010) Atomic force microscope-related study membrane-associated cytotoxicity in human pterygium fibroblasts induced by mitomycin C. J Phys Chem B 114:3833–3839

73. Radmacher M (2002) Measuring the elastic properties of living cells by the atomic force microscope. Methods Cell Biol 68:67–90
74. Flieger D, Renoth S, Beier I, Sauerbruch T, Schmidt-Wolf I (2000) Mechanism of cytotoxicity induced by chimeric mouse human monoclonal antibody IDEC-C2B8 in CD20 expressing lymphoma cell lines. Cell Immunol 204:55–63
75. Deans JP, Li H, Polyak MJ (2002) CD20 mediated apoptosis: signaling through lipid rafts. Immunology 107:176–182
76. Shan D, Ledbetter JA, Press OW (1998) Apoptosis of malignant human B cells by ligation of CD20 with monoclonal antibodies. Blood 91:1644–1652
77. Oflazoglu E, Audoly LP (2010) Evolution of anti-CD20 monoclonal antibody therapeutics in oncology. mAbs 2:14–19
78. Lingwood D, Simons K (2010) Lipid rafts as a membrane-organizing principle. Science 327:46–50
79. Bezombes C et al (2004) Rituximab antiproliferative effect in B-lymphoma cells is associated with acid-sphingomyelinase activation in raft microdomains. Blood 104:1166–1173
80. Unruh TL et al. (2005) Cholesterol depletion inhibits src family kinase-dependent calcium mobilization and apoptosis induced by rituximab crosslinking. Immunology 116:223–232
81. Janas E, Priest R, Wilde JI, White JH, Malhotra R (2005) Rituxan (anti-CD20 antibody)-induced translocation of CD20 into lipid rafts is crucial for calcium influx and apoptosis. Clin Exp Immunol 139:439–446
82. Kheirallah S et al (2010) Rituximab inhibits B-cell receptor signaling. Blood 115:985–994
83. Walshe CA et al (2008) Induction of cytosolic calcium flux by CD20 is dependent upon B cell antigen receptor signaling. J Biol Chem 283:16971–16984
84. Fletcher DA, Mullins RD (2010) Cell mechanics and the cytoskeleton. Nature 463:485–492
85. Shi X, Xu L, Yu J, Fang X (2009) Study of inhibition effect of Herceptin on interaction between Heregulin and ErbB receptors HER3/HER2 by single-molecule force spectroscopy. Exp Cell Res 315:2847–2855
86. Li M, Xiao X, Liu L, Xi N, Wang Y, Dong Z, Zhang W. Imaging and measuring the molecular force of lymphoma pathological cells using atomic force microscopy. Scanning (in press) DOI:10.1002/sca.21033
87. Wang H, Bash R, Yodh JG, Hager GL, Lohr D, Lindsay SM (2002) Glutaraldehyde modified mica: a new surface for atomic force microscopy of chromatin. Biophys J 83:3619–3625
88. Kada G, Kienberger F, Hinterdorfer P (2008) Atomic force microscopy in bionanotechnology. Nano Today 3:12–19
89. Muller DJ, Engel A, Amrein M (1997) Preparation techniques for the observation of native biological systems with the atomic force microscope. Biosens Bioelectron 12:867–877
90. Henderson RM, Schneider S, Li Q, Hornby D, White SY, Oberleithner H (1996) Imaging ROMK1 inwardly rectifying ATP-sensitive K^+-channel protein using atomic force microscopy. Proc Natl Acad Sci USA 93:8756–8760
91. Kirat KE, Burton I, Dupres V, Dufrene YF (2005) Sample preparation procedures for biological atomic force microscopy. J Microsc 218:199–207
92. Kada G et al (2001) Recognition force microscopy/spectroscopy of ion channels: applications to the skeletal muscle Ca^{2+} release channel (RYR1). Ultramicroscopy 86:129–137
93. Wang H, Kutner LO, Lin M, Huang Y, Grace MJ, Lindsay SM (2008) Imaging glycosylation. J Am Chem Soc 130:8154–8155
94. Li M, Liu L, Xi N, Wang Y, Dong Z, Li G, Xiao X, Zhang W (2011) Detecting the CD20 rituximab interaction forces using AFM single-molecule force spectroscopy. Chinese Sci Bull 56:3829–3835
95. Glennie MJ, French RR, Cragg MS, Taylor RP (2007) Mechanisms of killing by anti-CD20 monoclonal antibodies. Mol Immunol 44:3823–3837
96. Beers SA, Chan CHT, French RR, Cragg MS, Glennie MJ (2010) CD20 as a target for therapeutic type I and II monoclonal antibodies. Semin Hematol 47:107–114

Chapter 9
Cell Manipulation with Robot-Aided Optical Tweezers Technology

Songyu Hu, Youhua Tan and Dong Sun

Abstract Increasing demands for both accuracy and productivity in cell manipulation highlight the need for the use of robotics technologies. Optical tweezers are known for the ability to impose force and deformation on a micro-scaled object in a noninvasive manner. In this chapter, we introduce an integration of robotics technology into optical tweezers system for cell manipulation at single cell level, where optical tweezers function as special robot end effectors. Two particular cell manipulation researches are introduced: one is cell stretching for biomechanical property characterization, and the other is automated cell transportation for rapid and precise cell positioning in many biomedical applications. In cell stretching, a mechanical model of the optically stretched cell is developed, based on which cell stiffness can be extracted and biomechanical property can be characterized. Experiments are performed on human embryonic stem cells (hESC) and hESC-derived cardiomyocytes (hESC-CM) to clarify changes in mechano-membrane property during cardiac differentiation. In cell positioning, based on dynamics analysis of the trapped cell in motion, a closed-loop controller is designed for cell transportation, which is further verified by experimental tests on live cells. These approaches successfully demonstrate the effectiveness of the robot-aided optical tweezers technology in cell manipulation.

S. Hu · Y. Tan · D. Sun (✉)
Department of Mechanical and Biomedical Engineering,
City University of Hong Kong, Kowloon, Hong Kong
e-mail: medsun@cityu.edu.hk

Y. Guo (ed.), *Selected Topics in Micro/Nano-Robotics for Biomedical Applications*, 159
DOI: 10.1007/978-1-4419-8411-1_9, © Springer Science+Business Media New York 2013

9.1 Introduction

Nowadays, physical manipulation and characterization of cells has become important for understanding their biophysical properties in the normal and disease states. Recent studies on the pathophysiology of many human diseases have suggested that their pathogenesis might result from deviation in the structural and mechanical properties of cells as well as abnormal mechanotransduction [17, 18]. There may be connections among cellular structure, mechanical properties, biological function, and disease state [18, 33]. In other words, changes in microstructures of a cell may affect not only its mechanical properties but also the physiological functions. Abnormity of biological function might give rise to a disease. Many research studies have indicated that alterations of mechanical properties of cells may be associated with the onset and progression of some diseases. For example, osteoarthritic chondrocytes were found to exhibit significantly higher viscoelastic properties, such as equilibrium modulus, instantaneous modulus, and apparent viscosity compared with normal ones [39]. Airway smooth muscle cells obtained from rats with airway hyperresponsiveness showed greater stiffening, bigger contractile forces, and faster cytoskeleton remodeling than cells in control [1]. Mechanical properties varied for mouse fibroblasts and human breast epithelial cells from normal to cancerous and even metastatic state [9], and Young's modulus of cancerous human epithelial cells was one order of magnitude lower than normal cells [19]. Moreover, the membrane of plasmodium falciparum infected red blood cells (RBCs) became much stiffer, and subsequently the shear modulus of these infected RBCs was found to increase up to 10-fold during parasite development [33].

Cell manipulation consists of manipulating individual or large populations of cells in order to perform specific biological operating tasks, such as cell transportation, cell sorting or isolation, and cell stretching, etc. The rationale behind this is to characterize biophysical properties of cells at molecular level, and further, to provide clinical doctors with important information on the function of cells and organisms for regenerative medicine. There is a critical need to examine various physical and mechanical properties of cells through cell manipulation. Single cell study, in particular, has received increasing attention because of the ability to provide rich information about individual cells that is often overlooked in the analysis of cell population. A growing tendency to characterize the biomechanical property of living cells is to stimulate and then measure the responses to the *stimuli* of individual living cells. With the development of micromanipulation technologies, several experimental techniques have been developed for single cell manipulation and property characterization, such as micropipette aspiration [12], atomic force microscopy [29], cell poker [23], magnetic twisting cytometry [40], optical tweezers [3], optical stretcher [11], and microfabricated devices [20]. These methods have the ability to measure the mechanical responses with high accuracy even down to pico-Newton and nanometer for probing forces and cell deformations, respectively.

Over the past decades, robotics technologies have been demonstrated as proven tools in control, vision, and system intelligence for macro object manipulations, and have been extended to microscale objects via integration with MEMS technology. Benefiting from great advances of robotics in visual servoing [27], micro-force sensing and control [25], motion control and image processing [32], microrobotic manipulation of biological objects has attracted substantial attention. On the other hand, optical tweezers [3], which use highly focused low power laser beams to trap and manipulate particle on the order of pico-Newton (pN, 10^{-12} N) and nanometer (nm, 10^{-9} m) in a noncontact manipulation manner [28], have provided a revolutionary solution to manipulation and characterization of biological objects. An integration of robotics with optical tweezers into a new manipulation system will open up a new dimension of cellular biology that has substantial clinical and biological relevance.

This chapter will introduce our recent efforts in manipulating biological cells with robot-aided optical tweezers technologies. The following two cell manipulations will be reported particularly.

9.1.1 Cell Stretching

The cell stretching manipulation uses optical traps to deform the cell for measurement of elasticity of the cell. Through comparison between modeling and real test performance, Tan et al. [35–37] characterized biomechanics property of cell, and further, built biological link with measurement. It has been demonstrated in the literature that cell mechanics largely affects cellular functions and reflects the physiological and developmental state of cells [33]. Recent researches further suggested that the cell mechanical properties may be associated with the onset and progression of some diseases. For example, human red blood cells (RBCs) either infected by plasmodium falciparum or under various osmotic conditions exhibited different shear moduli [26, 35]. Metastatic cancer cells were much softer than their benign counterparts [5]. Cell transformation and tumorigenicity were associated with viscoelastic properties of cells [6]. Due to the unique role that stem cells play in tissue engineering and regenerative medicine, there exists an increasing demand for investigating the mechanics of stem cells during the process of differentiation and development [42].

9.1.2 Cell Positioning

Transportation and positioning of biological cells become increasingly important in contemporary biomedical researches. [15, 16] developed a closed-loop control algorithm based on dynamics analysis of the trapped cell in movement, which provided a novel solution for positioning cells rapidly and precisely, and exhibited

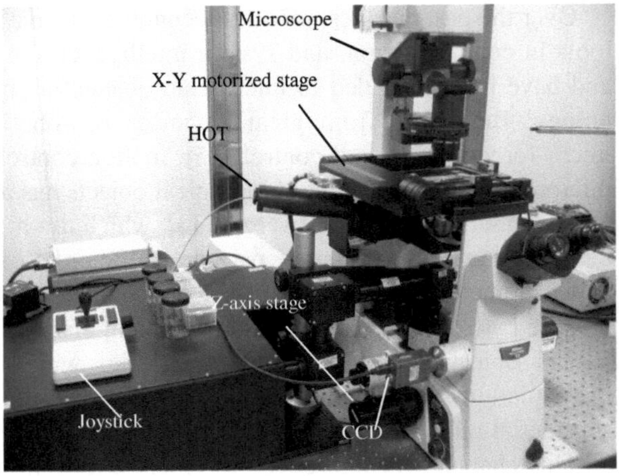

Fig. 9.1 Robotic manipulation system with optical tweezers

much biological relevance in cell-to-cell interaction, drug discovery, and tissue engineering. Some examples of cell transportation include: cell sorting and rapid cell arrays by combining microfluidics technology [4], which show significant potentials in drug discovery, tissue engineering, and fundamental cell biology [22, 43]; microsurgery of the living cells, where the cells need to be precisely positioned at the defined locations, and the laser microbeam is then used to ablate unicellular organisms, cells and subcellular structures (mitochondria, nucleolus, and chloroplast, etc.) for investigation of their corresponding functions [7]. Moreover, the rapid and precise transportation of cells can benefit the fabrication of pharmacophore–erythrocyte complexes [8], studies on cell communication [21], and cell fusion, etc.

In the following, after a brief introduction of the robot-tweezers manipulation system, details of the manipulation for cell stretching and cell positioning will be given respectively.

9.2 Robot-Tweezers Cell Manipulation System

Figure 9.1 shows the robotic manipulation system with optical tweezers (BioRyx 200, Arryx) in our laboratory, which mainly includes three modules: an executive module, a sensory module, and a control module. The *executive module* consists of an *X*–*Y*–*Z* motorized stage (ProScan, Prior Scientific) with microscopic objectives and a holographic optical trapping device (HOT) mounted on the *Z*-axis. Biological samples are contained in a home-built chamber, which is placed on the motorized stage with a positioning accuracy of 15 nm. The motions of optical traps can be controlled independently in three dimensions by the HOT device through modulating the phase distribution of a spatial light modulator (SLM). The *sensory module* includes an inverted microscope (Nikon TE2000, Japan), a CCD camera (FO124SC, Foculus), and a PCI image capture and processing card.

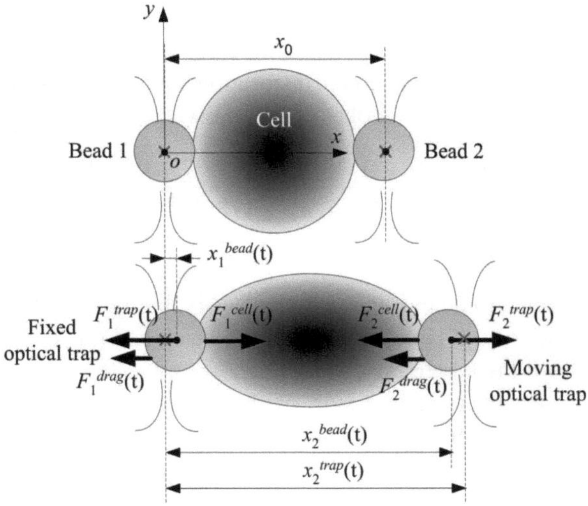

Fig. 9.2 Schematic of cell stretching, where $x_1^{\mathrm{bead}}(t)$, $x_2^{\mathrm{bead}}(t)$, and $x_2^{\mathrm{trap}}(t)$ are the positions of bead 1, bead 2, and the moving optical trap, and x_0 is the initial distance between the two beads. The *dots* and *crosses* denote the centroids of the beads and the centers of the traps

The positions of biological objects can be obtained through image processing, which serve as visual feedback to guide the manipulation process. The *control module* comprises a motion controller for the *X–Y–Z* motorized stage, a programmable phase modulator for the HOT device, and a host computer. All of the mechanical components were supported upon an anti-vibration table.

In the optical tweezers system, a continuous wave (CW) laser beam (YLM-3-1064-LP-AX1, IPG Photonics) with wavelength of 1,064 nm is sculpted by the HOT device and up to 200 optical traps can be created simultaneously. These traps serve as end effectors to manipulate multiple micro/nano-scaled objects with size ranged from 100 nm to 100 μm. A home-made operation interface is used for image processing and motion control of the optical traps and the motorized stage. A joystick controller is adopted for coarse positioning.

9.3 Cell Stretching

9.3.1 Modeling

Figure 9.2 illustrates a schematic of cell stretching with optical tweezers [38]. Two micro-beads are adhered to the cell surface and optical traps focus on the beads. Without loss of generality, we consider that two beads adhere to a cell in the diametrically opposite directions, where one bead is held in place with one trip, and the other bead was moved by the second trip along the joint line of centroids of the two beads. For simplicity, we define x coordinate direction to be parallel to the moving direction of the optical trap, such that the computation can be simplified in one dimension. The cell is stretched by progressively increasing the distance between the two beads.

Three forces applied to the bead were considered during cell stretching. These three forces are: viscous drag force exerted by the surrounding medium, optical trapping force, and cell restoration force due to elongation (Fig. 9.2).

According to Stokes' Law, the viscous drag force was calculated as

$$F^{\mathrm{drag}} = \frac{6\pi R\eta_0 V}{1 - 9/16(R/h) + 1/8(R/h)^3 - 45/256(R/h)^4 - 1/16(R/h)^5} = \beta V \tag{9.1}$$

where β is the drag coefficient, R is the radius of the trapped bead, η_0 is the fluid viscosity, V is the velocity of the bead, and h is the separation distance of the bead from the coverslip surface.

A general expression of the optical force on the trapped bead was approximately expressed as [24]

$$F^{\mathrm{trap}} = k\varepsilon = \begin{cases} k\varepsilon, & |\varepsilon| \le \varepsilon_0 \\ k\varepsilon \cdot e^{-\lambda(|\varepsilon| - \varepsilon_0)/\varepsilon_0}, & |\varepsilon| > \varepsilon_0 \end{cases} \tag{9.2}$$

where k is the trapping stiffness, ε denotes the derivation of the cell with respect to the laser focus, and ε_0 is the critical deviation beyond which the trapping force decays as an exponential function with coefficient λ. The critical deviation $\varepsilon_0$was measured from the calibration experiment, and the decay coefficient λ was determined by fitting the numerical results [30] with an exponential function.

The restoration force can be known through analysis of the dynamics of the two beads as follows:

$$m_1\ddot{x}_1^{\mathrm{bead}}(t) = F_1^{\mathrm{cell}} - F_1^{\mathrm{trap}} - F_1^{\mathrm{drag}} \tag{9.3}$$

$$m_2\ddot{x}_2^{\mathrm{bead}}(t) = F_2^{\mathrm{trap}} - F_2^{\mathrm{cell}} - F_2^{\mathrm{drag}} \tag{9.4}$$

where the subscripts 1 and 2 denote the corresponding responses of bead 1 and bead 2, respectively, m is the mass of the bead, $x_i^{\mathrm{bead}}(t)$ is the instantaneous position of the bead, and F_i^{cell} represents cell restoration force, which is the mechanical response of the cell.

The cell mechanical model developed in our previous work [34–36] is utilized to acquire the information of cell stiffness, and further, to characterize biomechanical properties of cells.

Biological cells especially suspended cells are usually simplified as a pressurized liquid drop enveloped by a spherical biomembrane [26, 37]. The resistance of a cell to stress is mainly dominated by its biomembrane. In our previous studies, we used membrane theory to model deformation behaviors of cells. The equilibrium equations were adopted to represent force balance of the cell membrane, expressed as follows

$$\frac{\partial T_m}{\partial \lambda_m}\lambda_m' + \frac{\partial T_m}{\partial \lambda_c}\lambda_c' = \frac{\rho'}{\rho}(T_c - T_m) \tag{9.5}$$

$$K_m T_m + K_c T_c = P \tag{9.6}$$

where λ_m and λ_c, T_m and T_c, and K_m and K_c are the principal stretch ratios, tensions, and curvatures, respectively. The prime denotes the derivative with respect to a specific angle. The indices m and c refer to the corresponding component in the meridian and circumferential directions of the deformed membrane, respectively. P is the turgor pressure acting on the membrane in the normal direction. The definition of all the quantities can be found in Tan [35].

9.3.2 Experiments

In cell stretching process, one bead is held by a fixed laser trap, while the other bead is driven to move by another trap. By progressively increasing the distance between the two beads, the cell can be stretched. The stretching direction is determined along the line passing the centroids of the two trapped beads. Based on the modeling results as described above, the cell motion and some biomechanical properties are determined, which can be compared to experimental data for characterization purpose.

The cells used in the tests were undifferentiated human embryonic stem cell (hESC) and hESC-derived cardiomyocytes (hESC-CM). Prior to experiments, it was necessary to attach microbeads to the cell surface. The treatment method was given elsewhere [35]. The mixture of cells and beads was contained in a home-built chamber. In experiments, the inner surface of the chamber was treated with 100 mg/ml bovine serum albumin (BSA, Sigma) to prevent beads and cells from sticking.

Figure 9.3 illustrates the stretching process on hESC. The positions of the beads were acquired by image processing. The viscous drag force calibration method was adopted to acquire the trapping stiffness k. Based on the cell model, the cell stiffness could be calculated. The critical value of the deviation distance ε_0 was obtained from calibration experiments. For example, when the cell was stretched to the maximum deformation, as shown in Fig. 9.3c, the cell restoration force was equal to the trapping force. Further, cell stretching then led to escape of the bead from the trap, as shown in Fig. 9.3d.

Prior to experiments, force calibrations were conducted to quantitatively obtain the relationship between the trapping force and the laser power. The viscous drag force calibration method was used in this study. We adopted the drag force to evaluate the optical force when the bead achieved equilibrium in the flow fluid. During experiments, the separation distance between the bead and the coverslip h was kept as 10 μm. The calibration results are given in Fig. 9.4.

Figures 9.5 and 9.6 illustrate the experimental data, where the results are averaged and the standard deviation is given. In the test, 23 hESC-CM cells and 18 H7 hESC cells were measured, respectively. The stretching force is calculated according to the relationship between laser powers and optical forces. By adjusting

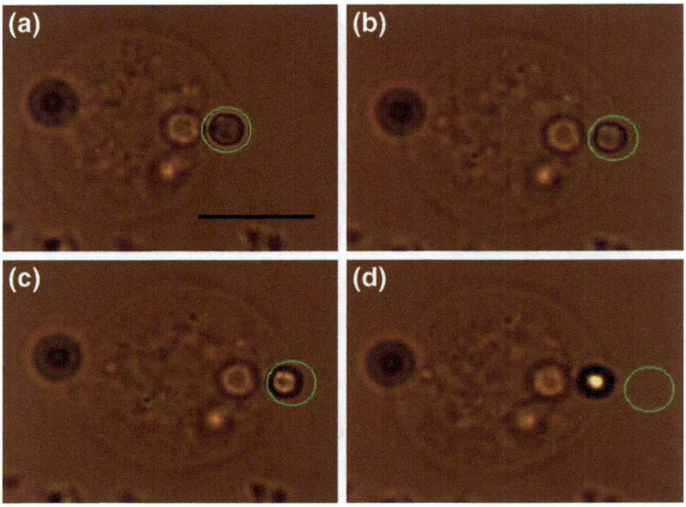

Fig. 9.3 Cell stretching process. **a** Before stretching, **b** during stretching, **c** stretched to the maximum deformation, **d** the bead escapes the trap. The scale bar is 10 μm

Fig. 9.4 Force calibration by trapping a polystyrene bead with radius of 1.55 μm at the separation depth $h = 10$ μm from the coverslip

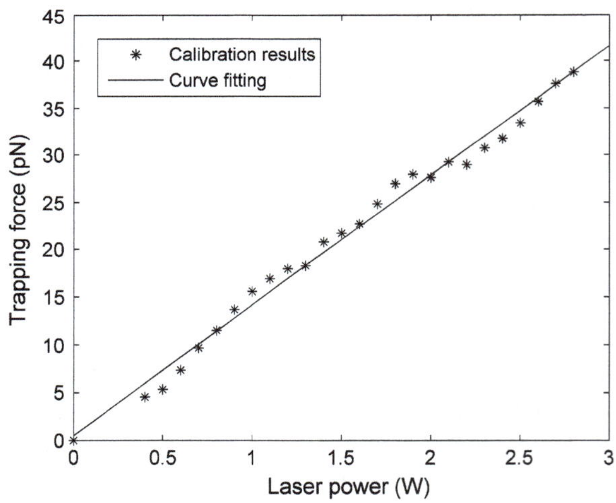

the power of the laser beam, different stretching forces can be obtained. Note that cell deformation was measured from the image at the moment when the cell was stretched to the maximum deformation. At each laser power, the cell deformation was measured. Then the relationship between the stretching force and the induced cell deformation could be established. According to Eqs. (9.1) through (9.4), the mechanical responses of the cell at both sides were calculated as shown in Fig. 9.5. The results show that the cell restoration force at the side of the moving trap (F_2^{cell}) is slightly larger than that at the side of the fixed trap (F_1^{cell}), which indicates that the resultant mechanical response ($F_2^{cell} - F_1^{cell}$) resists cell stretching induced by

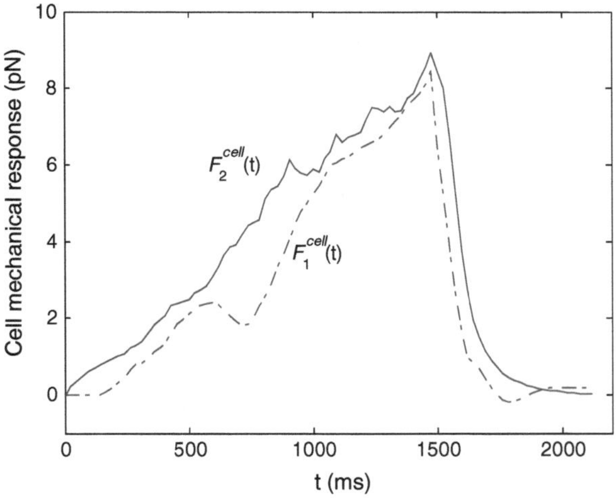

Fig. 9.5 Cell restoration forces, where $k = 7.8$ pN/μm, $\varepsilon_0 = 1.5\,\mu$m, and $\lambda = 1.47$

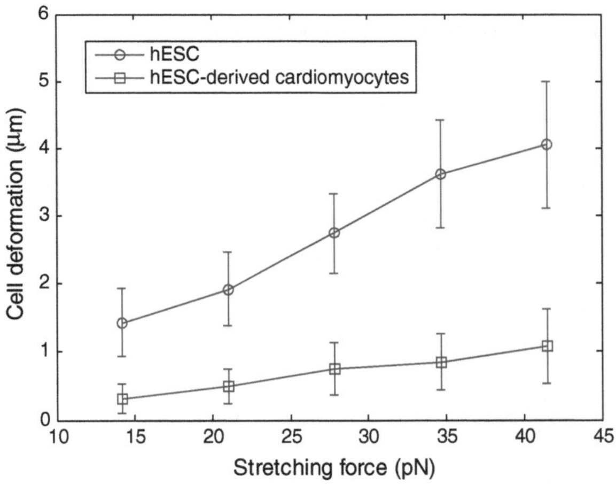

Fig. 9.6 Quantitative analysis of the experimental results for hESC with different differentiation potency. The *error bar* denotes the standard deviation of the results

the optical force. Figure 9.6 illustrates the quantitative analysis of the mechanical responses of hESC and hESC-CM in equilibrium during cell stretching. It is shown that the deformability of hESC cells is significantly larger than that of hESC-CM. To extract mechanical properties of cells, the cell mechanical model was applied to reveal the deformation behavior of cells in optical stretching. The modeling deformation-force relationship is quasi-linear, from which the cell stiffness k can be calculated as 12.5 ± 2.4 μN/m and $4.34 + 0.91$ μN/m for hESC-CM and hESC, respectively. This indicates that the cell stiffness of hESC-CM is about 3-fold of that of hESC. Further, the area compressibility modulus was characterized as 0.36 ± 0.092 N/m and 1.95 ± 0.24 N/m (mean ± standard deviation) for hESC and hESC-CM, respectively. This result also implies that differentiated hESC-CM are much stiffer than the undifferentiated hESC, suggesting that cell mechanics varies during hESC differentiation.

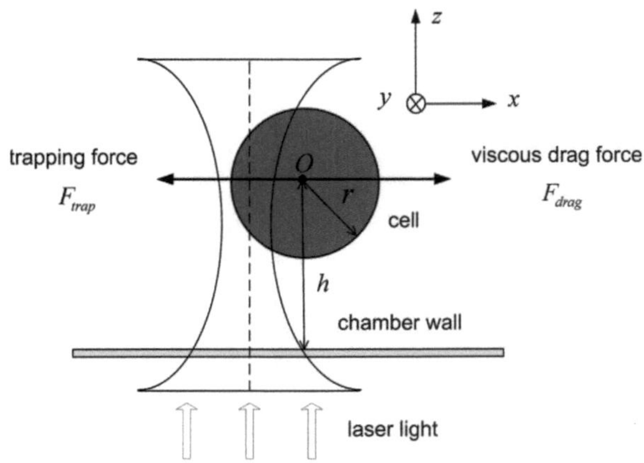

Fig. 9.7 A trapped cell in movement

9.4 Cell Positioning

9.4.1 Modeling and Control

We now introduce how to use optical trap to transport a cell. When a trapped cell moves in the liquid, it is exerted by the trapping force and the viscous drag force simultaneously. Figure 9.7 shows the forces applied to the trapped cell during its movement in X–Y plane. The drag force causes the cell to deviate from the center of the trap in the lateral direction. The trapping force, generated by the laser beam due to the change of the angles of the incidence and refraction, pulls the cell into alignment with the beam's central axis. The equation of the motion of the cell in the lateral plane is

$$m\ddot{q} = F^{\mathrm{trap}} - F^{\mathrm{drag}} \tag{9.7}$$

where m is the cell's mass, $q \in R^{2\times1}$ is the position of the cell in the coordinated frame. Based on the works reported by Hertz [11] and Ashkin [2], we can get the numerical solution of the force that is a function of the offset between the trap center and the cell center, as detailed in Hu et al. [14]. Since the effect of the inertia force can be ignored in low Reynolds number environment [13], Eq. (9.7) can be simplified as [41]:

$$0 = F^{\mathrm{trap}} - F^{\mathrm{drag}} \tag{9.8}$$

Define $I \in R^{2\times1}$ as the trap position. Then, the deviation or offset of the cell from the center of the trap can be represented as $\varepsilon = \|I - q\|$. It has been concluded in the literature that the trapping force of the cell is approximately proportional to the displacement of the cell from the focus when the offset is not larger than the critical deviation ε_0, in the format

$$F^{\mathrm{trap}} = k(I - q), \quad \|I - q\| \le \varepsilon_0 \tag{9.9}$$

where k is the stiffness of the optical trap as defined in (9.2). When the offset exceeds ε_0, the trapping force decreases as the offset increases, and becomes zero when the offset is very large. Therefore, it is ideal to confine the displacement to be less than ε_0 in the cell transportation.

The drag force applied to a moving cell in liquid can be calculated by Faxen's law [11] as follows:

$$F^{\text{drag}} = \beta\dot{q} \tag{9.10}$$

where β is the drag coefficient as defined in (9.1).

Substituting (9.9) and (9.10) into (9.8) yields the dynamics equation of the motion of the cell in optical trap, expressed as

$$\frac{\beta}{k}\dot{q} = I - q, \|I - q\| \le \varepsilon_0 \tag{9.11}$$

Based on the dynamics Eq. (9.11), we can design a controller for the trap position input in the following format:

$$I - q = \begin{cases} c, & \|I - q\| \le \varepsilon_0 \\ \frac{I-q}{\|I-q\|}\varepsilon_0, & \|I - q\| > \varepsilon_0 \end{cases} \tag{9.12}$$

where c is a closed-loop controller. Based on the dynamics (9.11), c is designed to be feedforward plus PI-type control [15, 16], i.e.,

$$c = \frac{\beta}{k}\dot{q}^d + K_p e + K_i \int_0^t e d\xi \tag{9.13}$$

where $K_p$and K_i are positive feedback control gains, ξ is a time variable from 0 to t, and e is the position error between the desired cell position q^d and the actual cell position q, given by

$$e = q^d - q \tag{9.14}$$

Substituting the controller (9.12) – (9.13) into (9.11), we obtain the following closed-loop dynamics when $\|I - q\| \le \varepsilon_0$,

$$\frac{\beta}{k}\dot{e} + K_p e + K_i \int_0^t e d\xi = 0 \tag{9.15}$$

Equation (9.15) implies that $e \to 0$ as time $t \to \infty$.

Utilizing the above controller, the cell can move along with the desired trajectory as long as the condition $\|I - q\| \le \varepsilon_0$ is satisfied.

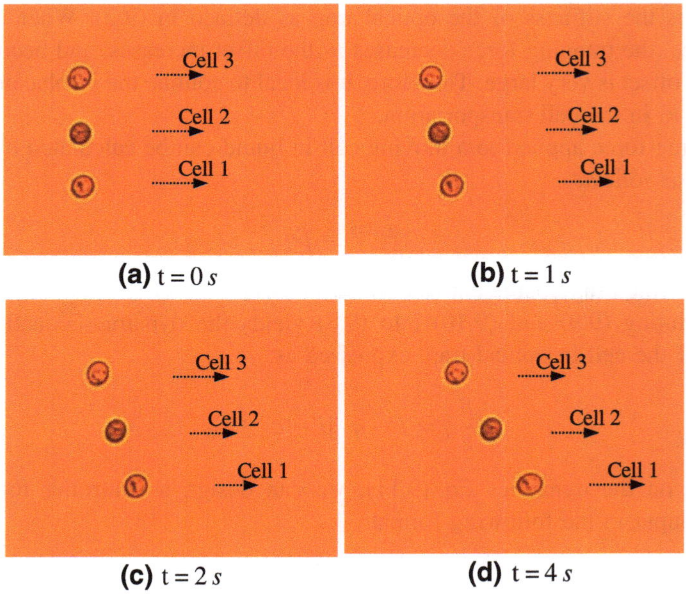

Fig. 9.8 Process of transportation of three yeast cells with different velocities

Fig. 9.9 Position tracking performance of cell 1

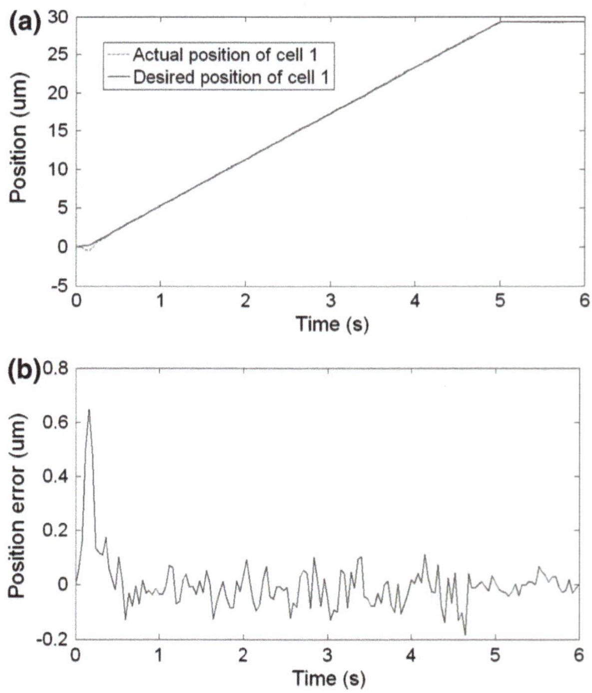

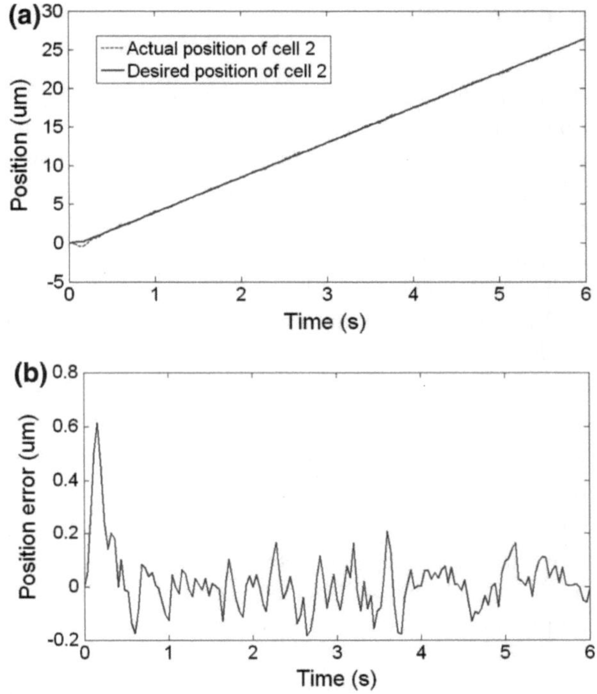

Fig. 9.10 Position tracking performance of cell 2

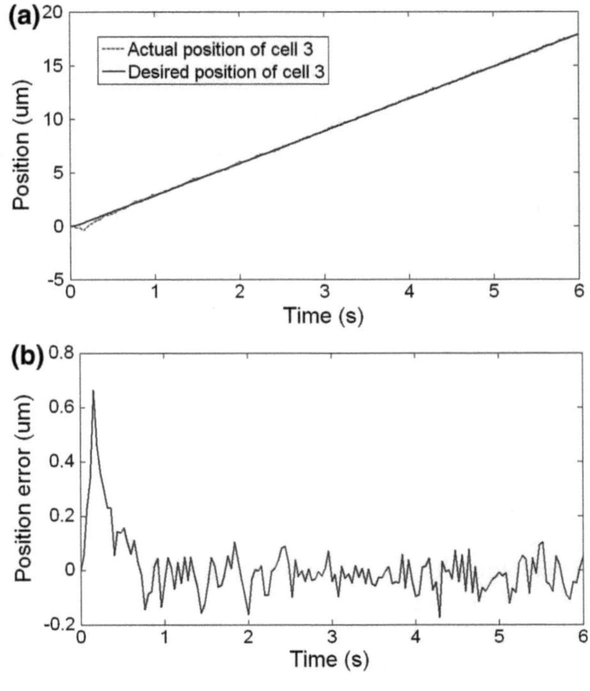

Fig. 9.11 Position tracking performance of cell 3

9.4.2 Experiments

To verify the effectiveness of the proposed approach, experiments were conducted on transporting the cells with the robot-tweezers manipulation system, as shown in Fig. 9.1. Considering the fact that few calculations or measurements of the full stiffness matrix of non-spherical objects trapped in optical tweezers have been reported in the literature (i.e., [31]), we mainly discussed the transportation of spherical cells in the study. In our experiments, the yeast cells were used. Since the radius of the yeast cell to be used in our experiments is about 5 μm, ε_0 was defined to 2.2 μm. The sampling frequency of the control system, which mainly depends on the rate of computing phase hologram and updating SLM, was set as 20 Hz. All the experiments were performed at 37 °C in 5 % CO2 for cell viability.

The experiment was to control cells to move along desired trajectory. In order to choose appropriate parameters for the controller, $\frac{\beta}{k}$ should be identified first. In this study, $\frac{\beta}{k}$ was determined using the viscous drag force method [27], which was 0.09 s, when the laser power is 0.2 w. It is worth noting that in order to precisely move the cell along the desired trajectory under the constraint $\|I - q\| \leq \varepsilon_0$, the motion should be well planed according to the stiffness of the optical trap, and properties of the cell and the suspending medium. If the planned motion velocity is too high, the cell fails to follow. In the experiment, we calibrated the maximum moving velocity in moving along straight line, which was 15 μm/s, with the sampling frequency set at 20 Hz.

In the experiment, three yeast cells were transported at different velocities simultaneously, with 6, 4.5, and 3 μm/s, respectively. The control gains were $K_p = \text{diag}\{0.4, 0.4\}$, and $K_i = \text{diag}\{0.2, 0.2\}$. Figure 9.8 shows the transportation process. In the figures, the dashed lines and the arrows represent the trajectories and the moving directions of the cells. Figures 9.8b–d show the different positions of the three yeast cells, with different velocities. Figure 9.9 through Figs. 9.10 and 9.11 illustrate the position tracking performance of the three yeast cells in the tracking control. It is seen that all of them exhibit good position accuracy.

9.5 Conclusions

This chapter presents a robot-aided optical tweezers technology for cell manipulation at single cell level. Two cell manipulation approaches have been introduced. One is to use optical traps to stretch cell and measure cell stiffness property based on the developed cell model. Through comparison between modeling and experimental results, the cell biomechanics property can be characterized. The other one is to use optical trap to transport cell. Upon analysis of cell dynamics model in movement, a closed-loop controller is developed for positioning cell rapidly and precisely. Cell stretching and cell transportation represent two typical cell manipulation operations, and exhibit great potential in biomedical applications.

The other types of cell manipulation, such as cell fusion, cell sorting, and cell adhesion etc., will be further investigated and reported in the future.

References

1. An SS, Fabry B, Trepat X, Wang N, Fredberg JJ (2006) Do biophysical properties of the airway smooth muscle in culture predict airway hyperresponsiveness? Am J Respir Cell Mol Biol 35:55–64
2. Ashkin A (1992) Forces of a single-beam gradient laser trap on a dielectric sphere in the ray optics regime. Biophys J 61(2):569–582
3. Ashkin A, Dziedzic JM (1987) Optical trapping and manipulation of single cells using infrared laser beams. Nature 330:769–771
4. Applegate RW, Squier J, Vestad T, Oakey J, Marr DWM, Bado P, Dugan MA, Said AA (2006) Microfluidc sorting system based on optical waveguide integration and diode laser bar trapping. Lab Chip 6:422–426
5. Cross SE, Jin YS, Rao JY, Gimzewski JK (2007) Nanomechanical analysis of cells from cancer patients. Nat Nanotechnol 2:780–783
6. Darling EM, Zauscher S, Block JA, Guilak F (2007) A thin-layer model for viscoelastic, stress-relaxation testing of cells using atomic force microscopy: do cell properties reflect metastatic potential? Biophys J 92:1784–1791
7. Greulich KO (1999) Micromanipulation by light in biology and medicine: the laser microbeam and optical tweezers. Birkhäuser, Berlin
8. Grover SC, Skirtzch AG, Gauthier RC, Grover CP (2001) Automated single-cell sorting system based on optical trapping. J Biomed Opt 6(1):14–22
9. Guck J, Schinkinger S, Lincoln B, Wottawah F, Ebert S, Romeyke M, Lenz D, Erickson HM, Ananthakrishnan R, Mitchell D, Kas J, Ulvick S, Bilby C (2005) Optical deformability as an inherent cell marker for testing malignant transformation and metastatic competence. Biophys J 88:3689–3698
10. Guck J, Ananthakrishnan R, Mahmood H, Moon TJ, Cunningham CC, Kas J (2001) The optical stretcher: a novel laser tool to micromanipulate cells. Biophys J 81:767–784
11. Hertz HM (1995) Standing-wave acoustic trap for nonintrusive positioning of microparticles. J Appl Phys 78(8):4845–4849
12. Hochmuth RM (2000) Micropipette aspiration of living cells. J Biomech 33:15–22
13. Hu Z, Wang J, Liang J (2007) Experimental measurement and analysis of the optical trapping force acting on a yeast cell with a lensed optical fiber probe. Opt Laser Technol 39:475–480
14. Hu S, Sun D, Feng G (2010) Dynamics analysis and closed-loop control of biological cells in transportation using robotic manipulation system with optical tweezers. In: IEEE international conference on automation science and engineering, Toronto, pp 240–245
15. Hu S, Sun D (2011a) Automated transportation of single cells using robot-tweezer manipulation system. Journal of Laboratory Automation 16(4):263–270
16. Hu S, Sun D (2011b) Automatic transportation of biological cells with robot-tweezer manipulation system. International Journal of Robotics Research 30(14):1681–1694
17. Ingber DE (2003) Mechanobiology and diseases of mechanotransduction. Ann Med 35:564–577
18. Lee GH, Lim CT (2007) Biomechanics approaches to studying human diseases. Trend Biotechnol 25:111–118
19. Lekka M, Laidler P, Gil D, Lekki J, Stachura Z, Hrynkiewicz AZ (1999) Elasticity of normal and cancerous human bladder cells studied by scanning force microscopy. Eur Biophys J 28:312–316
20. Loh O, Vaziri A, Espinosa HD (2009) The potential of MEMS for advancing experiments and modeling in cell mechanics. Exp Mech 49:105–204

21. Losert W, Poole C, Skupsky R (2005) Investigating gradient sensing in cells through optical micromanipulation. In: Proceeding of the SPIE, vol 5699, pp 281–287
22. Lu Z, Moraes C, Zhao Y, You L, Simmons CA, Sun Y (2010) A micromanipulation system for single cell deposition. In: IEEE international conference on robotics and automation, Alaska, pp 494–499
23. McConnaughey WB, Petersen NO (1980) Cell poker: an apparatus for stress-strain measurements on living cells. Rev Sci Instrum 51:575–580
24. Neuman KC, Block SM (2004) Optical trapping. Rev Sci Instrum 75:2787–2809
25. Mencaissi A, Eisiberg A, Carrozza MC, Dario P (2003) Force sensing microinstrument for measuring tissue properties and pulse in microsurgery. IEEE/ASME Trans Mechatron 8(1):10–17
26. Mills JP, Diez-Silva M, Quinn DJ, Dao M, Lang MJ, Tan KSW, Lim CT, Milon G, David PH, Mercereau-Puijalon O, Bonnefoy S, Suresh S (2007) Effect of plasmodial RESA protein on deformability of human red blood cells harboring Plasmodium falciparum. Proc Natl Acad Sci USA 104:9213–9217
27. Nelson B, Yu Z, Vikramaditya B (1998) Sensor-based microassembly of hybrid MEMS devices. IEEE Control Syst Mag 18(6):35–45
28. Pine J, Chow G (2009) Moving live dissociated neurons with an optical tweezer. IEEE Trans Biomed Eng 56(4):1184–1188
29. Radmacher M (1997) Measuring the elastic properties of biological samples with the AFM. IEEE Eng Med Biol Mag 16:47–57
30. Simmons RM, Finer JT, Chu S, Spudich JA (1996) Quantitative measurements of force and displacement using an optical trap. Biophys J 70:1813–1822
31. Simpson SH, Hanna S (2010) Holographic optical trapping of microrods and nanowires. J Opt Soc Am A 27(6):1255–1264
32. Sun Y, Nelson BJ (2002) Biological cell injection using an autonomous microrobotic system. Int J Robot Res 21:861–868
33. Suresh S, Spatz J, Mills JP, Micoulet A, Dao M, Lim CT, Beil M, Seufferlein T (2005) Connections between single-cell biomechanics and human disease states: gastrointestinal cancer and malaria. Acta Biomater 1:15–30
34. Tan Y, Sun D, Huang W, Cheng SH (2008) Mechanical modeling of biological cells in microinjection. IEEE Trans Nanobiosci 7(4):257–266
35. Tan Y, Sun D, Wang J, Huang W (2010) Mechanical characterization of human red blood cells under different osmotic conditions by robotic manipulation with optical tweezers. IEEE Trans Biomed Eng 57(7):1816–1825
36. Tan Y, Sun D, Huang W (2010) Mechanical modeling of red blood cells during optical stretching. J Biomech Eng Trans ASME 132:044504
37. Tan Y, Sun D, Huang W, Cheng SH (2010) Characterizing mechanical properties of biological cells by microinjection. IEEE Trans Nanobiosci 9(3):171–180
38. Tan Y, Sun D, Cheng SH, Li R (2011) Robotic manipulation with optical tweezers in cell stretching process for biomechanical characterization. In: IEEE international conference on robotics and automation, Shanghai
39. Trickey WR, Lee GM, Guilak F (2000) Viscoelastic properties of chondrocytes from normal and osteoarthritic human cartilage. J Orthop Res 18:891–898
40. Wang N, Ingber DE (1995) Probing transmembrane mechanical coupling and cytomechanics using magnetic twisting cytometry. Bioche Cell Biol 73:327–335
41. Wu Y, Sun D, Huang W (2011) Mechanical force characterization in manipulating live cells with optical tweezers. J Biomech 44:741–746
42. Yu H, Tay CY, Leong WS, Tan SCW, Liao K, Tan LP (2010) Mechanical behavior of human mesenchymal stem cells during adipogenic and osteogenic differentiation. Biochem Biophys Res Commun 393:150–155
43. Zahn M, Renken J, Seeger S (1999) Fluorimetric multiparameter cell assay at the single cell level fabricated by optical tweezers. FEBS Lett 443:337–340

Chapter 10
A Novel Robot-Assisted Catheter Surgery System with Force Feedback

Shuxiang Guo, Nan Xiao and Baofeng Gao

Abstract Endovascular intervention is expected to become increasingly popular in medical practice, both for diagnosis and for surgery. Accordingly, researches of robotic systems for endovascular surgery assistant have been carried out widely. Robotic system takes advantages of higher precision, can be controlled remotely, etc. In this chapter, a novel robotic catheter manipulation system is presented. The developed system consists of two parts, one is the controller and the other one is the catheter manipulator. The controller is designed to simulate the surgeon's operating procedure, and the catheter manipulator takes the same movement motion with the controller. An Internet-based communication between the controller and the catheter manipulator has been built. Performance evaluation of the mechanical system was assessed by experiments to quantify the precision of both controller and catheter manipulator. In order to keep the safety of the operation, force monitoring system are developed and installed on the system. Finally, two-way remote surgery experiments were carried out between Japan and China.

10.1 Background

Intracavity intervention is expected to become increasingly popular in the medical practice, both for diagnosis and for surgery. A lot of diagnosis and medical surgery with an endoscope or a catheter are performed for minimally invasive surgery recently. There are a lot of advantages as earliness, etc. However, it requires a lot of skills for the operation so that this may do the operation in the inside of the body

S. Guo (✉) · N. Xiao · B. Gao
Kagawa University, Kagawa, Japan
e-mail: guo@eng.kagawa-u.ac.jp
URL: http://www.guolab.org/

Y. Guo (ed.), *Selected Topics in Micro/Nano-Robotics for Biomedical Applications*,
DOI: 10.1007/978-1-4419-8411-1_10, © Springer Science+Business Media New York 2013

that cannot be watched directly. The intravascular neurosurgery brings a lot of challenges to doctors. Much more skills and experience are required to insert the catheter. Any mistakes would hurt patients and cause damages. During the operation X-ray camera is used, long time operation will cause damage to the patient. Although doctors wear protecting suits, it is very difficult to protect doctors' hands and faces from the radiation of the X-ray. In intravascular neurosurgery, the catheter is inserted through patients' blood vessel. An experienced neurosurgeon can achieve a precision of about 2 mm in the surgery. However, the contact force between the blood vessel and the catheter cannot be sensed. There are dangers of mingling or breaking the blood vessels. To overcome these challenges, we need better techniques and mechanisms to train doctors.

The aim of this project was to develop a new catheter operating system for intravascular neurosurgery. Such surgery presents many challenges:

1. Doctors must be very well trained and possess the skills and experience to insert catheters. Intravascular neurosurgery is much more difficult than traditional surgery and there are only few skilled doctors who can perform this type of operation. To keep pace with the growing number of patients, a mechanism is required to allow the training of sufficient numbers of doctors.
2. During the operation, doctors check the position of the catheter tip using the X-ray camera. Although they wear protective suits, it is very difficult to shield the doctor's hands and face from the effects of the X-ray radiation, which may result in radiation-related illness after long periods of exposure.
3. In intravascular neurosurgery, catheters are inserted into the patient's blood vessels, which in the brain are very sensitive. When operating in this area, extreme care is required to avoid damaging the fragile vessels. An experienced neurosurgeon can achieve an accuracy of about 2 mm. However, as the force of contact between the blood vessel and the catheter cannot be judged accurately by the doctor, catheters are obviously not suited for this type of surgery.
4. Sometimes doctors cannot be physically present to operate on patients. Therefore, Internet-based master–slave systems are required for such cases so that the operation can proceed.

For these reasons, a remote-controlled catheter system is urgently required.

One of the most popular product is a robotic catheter placement system called Sensei Robotic Catheter System [1–3] offered by Hansen Medical. The Sensei provides the physician with more stability and more force in catheter placement with the Artisan sheath compared to manual techniques, allowing for more precise manipulation with less radiation exposure to the doctor, commensurate with higher procedural complications to the patient. Because of the sheath's multiple degrees of freedom, force detection at the distal tip is very hard.

Catheter Robotics Inc. has developed a remote catheter system called Amigo. This system has a robotic sheath to steer catheters which is controlled at a nearby work station, in a manner similar to the Sensei system. The first use of this system in human was in April 2010 in Leicester UK, where it was used to ablate atrial flutter [4].

Magnatecs Inc. has produced their 'Catheter Guidance Control and Imaging' (CGCI) system, which has four large magnets placed around the table, with customized catheters containing magnets in the tip. The catheter is again moved by the magnetic fields and is controlled at a nearby work station.

The Stereotaxis Inc. developed a magnetic navigation system: the Stereotaxis Niobe II [5]. The system facilitates precise vector-based navigation of magnetically enabled guide wires for percutaneous coronary intervention (PCI) by using two permanent magnets located on opposite sides of the patient table to produce a controllable magnetic field.

Although these products have been developed and marketed, most concern is still the safety of the system. Force information of the catheter during the operation is very important to ensure the safety of the surgery. However, detection of the force on catheters is very hard to solve in these systems. A potential problem with a remote catheter control system is the lack of mechanical feedback that one would receive from manually controlling a catheter [1, 6–11]. To solve the problems, in this chapter we proposed a novel catheter system with force feedback and monitor.

On the other hand, the catheter is required to insert into narrow blood vessel, such as in neurosurgery. From these products and researches, it can be found that active catheters are used. Sizes of these active catheters are too large to fulfill many neurosurgeries. Therefore, a catheter manipulation system which can manipulate normal catheter is in urgent demanded.

Remote surgery methods are reported. In most cases, these methods are considered as basic researches. It has a long way to go to use these technologies to the actual surgeries. However, remote surgery takes many advantages. In this chapter, a remote control is built.

10.2 Design of the Robot-Assisted Catheter Surgery System

The robot-assisted catheter surgery system is designed to consist of two parts; one is the controller of the system which is to be placed at the doctor side. The other part is the catheter manipulator which is to be placed at the patient side. An Internet-based remote control is designed between these two parts. Figure 10.1 shows the developed system.

This is a master–slave structure system. The controller of the system is the master part and the catheter manipulator is the slave part. The movement of slave side should follow the master side, it means the movable parts of controller and manipulator keep the same displacement, speed, and rotation angle; therefore, the surgeon could operate the system smoothly and easily.

10.2.1 Design of the Master Machine

Experienced physicians tend to do the operation with their familiar way. Therefore, the controller is designed to simulate the motion of catheter during the operation

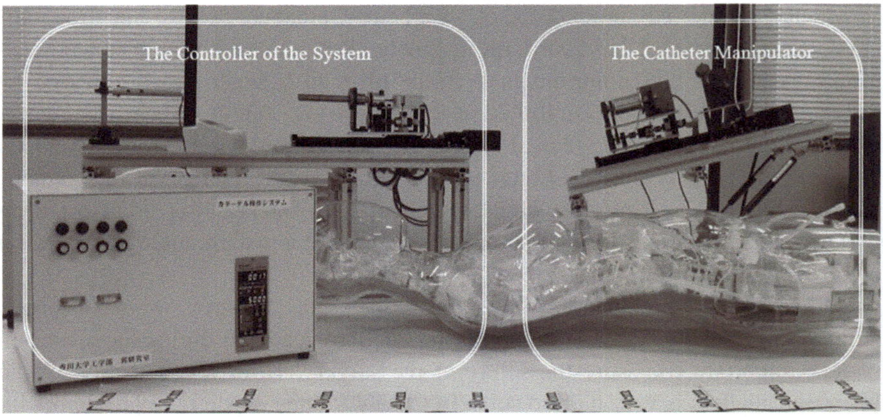

Fig. 10.1 The robot-assisted catheter manipulation system

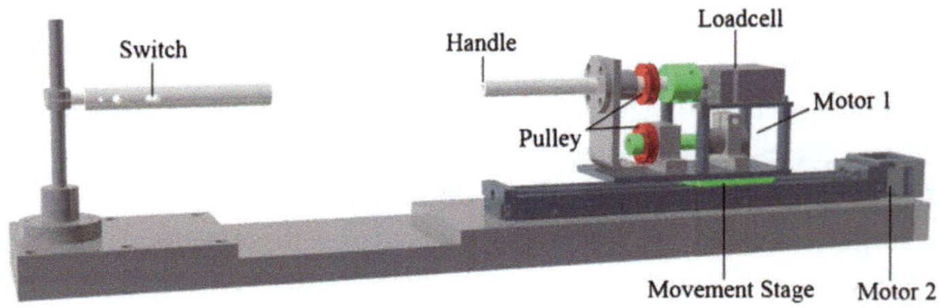

Fig. 10.2 Controller of the system-master machine

procedure. The motion of the catheter could be divided into two kinds, movement along the axial direction of the catheter and rotation about the radial direction of the catheter. Therefore, the controller is designed following these concepts.

Figure 10.2 shows the controller of the system. The controller is the master side of the whole system and is placed in the surgeon side. Surgeons carry out operations by using the controller. A switch placed on the left handle is used to control these two graspers in catheter manipulator side; only one switch is enough because the catheter is clamped by one grasper at the same time. Operators' action is detected by using the right handle. The movement part of catheter manipulator keeps the same motion with the right handle of the controller. The right handle can measure two actions of the surgeon's hand, one is axial movement and the other one is radial movement. The handle is sustained by a bearing, and is linked to a load cell; a pulley is fixed on the handle. A dc motor (Motor 1) with encoder is applied to generate torque feedback. A pulley which is couple to the upper one is fixed to the axle of the motor. All these parts are placed on a movement stage driven by a stepping motor (Motor 2).

Measurement of the axial movement is realized as following. A pulling/pushing force is measured by the loadcell when the operator pull or push the handle, according to this pulling force, the movement output displacement to keep the

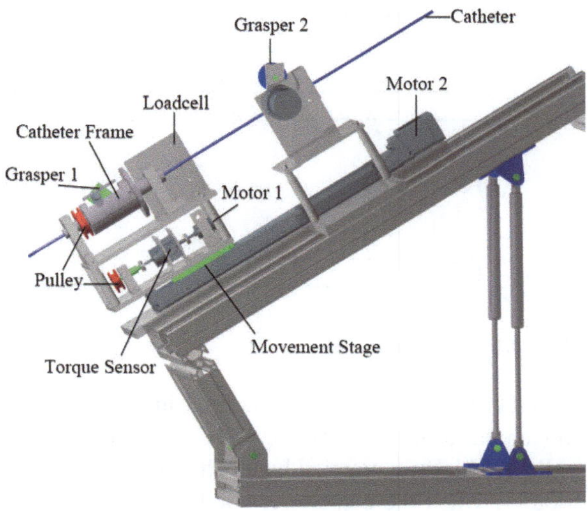

Fig. 10.3 Catheter manipulator-slave machine

handle following the surgeon's hand. Force feedback can be displaced by adjusting moving speed of the movement stage. The displacement and speed of the movement stage are sent to the catheter manipulator side, then the catheter manipulator keep synchronization with the controller. When the operator rotates the handle, the rotation angle is measured by an encoder installed in the dc motor. The dc motor is working in the current control mode to generate the damping to the surgeon. The damping is calculated by the torque information from the catheter manipulator side.

10.2.2 Design of the Slave Machine of the System

The structure of the catheter manipulator is also the controller; this means that the catheter manipulator could keep the same motion with the operator's hand. The operation will become visualized and easy to begin. On the other hand, this structure can realize the mechanical feedback to the surgeon.

Figure 10.3 shows the catheter manipulator. This part is placed in the patient side. The catheter is inserted by using this mechanism. This part contains two DOFs, one is axial movement along the frame, and the other one is radial movement. Two graspers are placed at this part. The surgeon can drive the catheter to move along both axial and radial when the catheter is clamped by grasper 1. The catheter keeps its position and the catheter-driven part can move freely when the catheter is clamped by grasper 2. Inserting motion of the catheter is as shown in Fig. 10.4.

To realize axial movement, all catheter-driven part are placed and fixed on a movement stage (the green plate under motor 1 in Fig. 10.3). The movement stage is driven by a screw which is driven by a stepping motor (motor 2 in Fig. 10.3). On

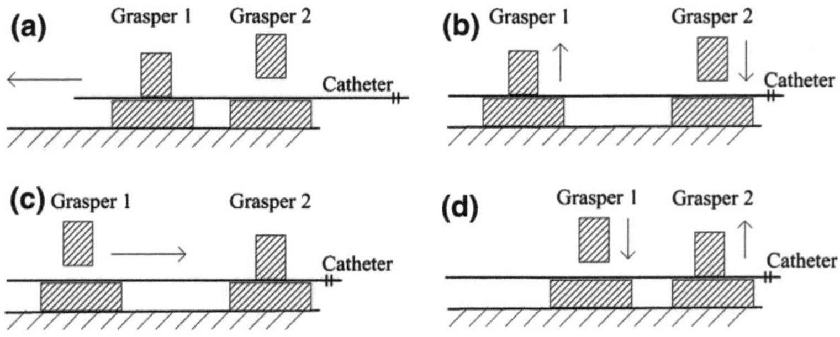

Fig. 10.4 Insertion process

the other hand, a DC motor (motor 1 in Fig. 10.3) is employed to realize the radial movement of the catheter. The DC motor is coupled to the catheter frame by two pulleys which are coupled by a belt with teeth. The catheter is driven to rotating by motor 1 when the catheter is fixed on the frame by grasper 1.

Torque sensor is applied in this system to measure the torque information during the operating. The torque information will be send to the controller side and generate a torque feedback to the surgeon. The torque sensor is linked to motor 1 and the axle of the pulley below. The resisting torque of the catheter can be transmitted to the torque sensor by coupled pulleys then measured by the torque sensor.

10.2.3 Communication Between Master Machine and Slave Machine

An Internet-based remote control system is built. Figure 10.5 shows the schematic diagram of the communication. A server-client structure is used to realize the communication. In the communication system, server could be built in each side of the robotic catheter manipulation system. Two kinds of data are transmitted between server and client; one is the control data between master and slave. In this part, rotation degrees of the handle and moving displacement of the movement stage will be sent to the slave side from master side. At the same time, rotation degrees of the catheter, moving displacement of the movement stage, force data from the loadcell, and torque data from the torque sensor will be sent to the master side. The other kind of data is image data. Compared with control data, the amount of image data is very large. To keep the safety of the operation, these two kinds of data are transmitted separately. IP camera is employed to get and transmit the image.

10.2.4 Force Feedback in the System

It is very important to displace haptic feedback to surgeons, however, to realize the haptic feedback is rather difficult. In the developed system, a force feedback structure is designed.

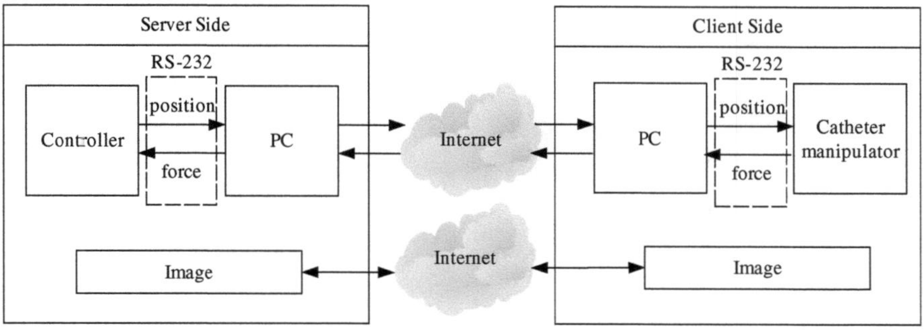

Fig. 10.5 Schematic diagram of the communication

Fig. 10.6 Force measurement mechanism

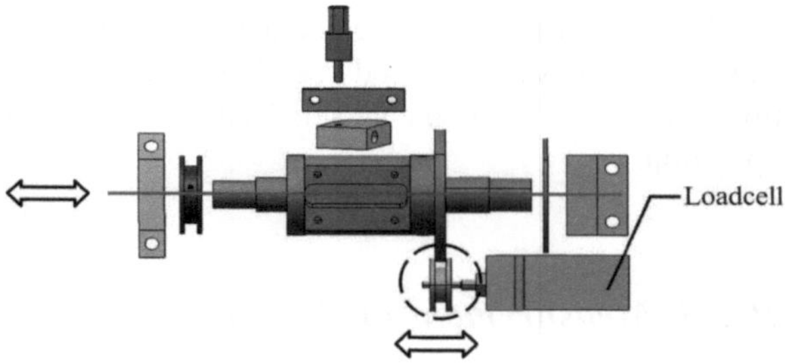

Resisting force acting on the catheter can be measured and will be sent to the controller and generated a haptic feedback to the surgeon. To measure the resisting force a mechanism is designed as shown in detail in Fig. 10.6. A loadcell which is fixed on the movement stage is employed to measure the resisting force. A clamp plate fixed on the loadcell is linked to the catheter frame which is supported by two bearings. The resisting force acting on the catheter in the axial direction can be detected by the loadcell when the catheter is fixed on the frame. The clamp plate does not affect the rotating motion of the catheter frame.

10.3 Real-Time Force Monitoring

During the operation of intravascular neurosurgery, it is significant to obtain the contact force information between catheter and blood vessel [12–15]. In order to detect the contact force information between catheter and blood vessel, we developed an intelligent force sensor system for robotic catheter systems. By using the developed force sensors system, we can obtain the contact force information and feedback it to the surgeon. If there are no force sensors on the catheter, it is easy to damage the blood vessel during operating, because the blood vessel is fragile. Figure 10.7 shows the comparison of safety between without force sensors on catheter and with force sensors on catheter.

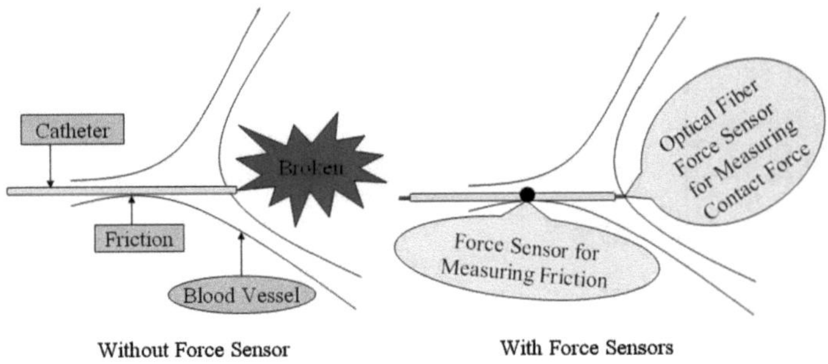

Fig. 10.7 Comparison of safety between two situations (without force sensors and with force sensors)

There are two kinds of the contact force. One is generated when the side of the catheter contacts to the blood vessel, and the other one is generated when the end-tip contacts to the blood vessel. Both of them could cause damage during the operation.

10.3.1 Measurement of the Side Contact Force

To measure the contact force in the side of the catheter, a novel force sensor is developed. The dimension of the force sensor is 4 × 2 × 0.5 mm. Three sensors are installed in the catheter in this system. The sensors are made of the sensitive material. Three rubber sensors are fixed on the catheter in a line as shown in Fig. 10.8. Calibration experiments were carried out, calibration results and the curves were fitted by MATLAB curve fitting tool box. Cubic polynomial which is described with Eq. 10.1 is adopted to describe the curve and the coefficients of the equation are given in Table 10.1.

10.3.2 Measurement of the End-Tip Contact Force

With tip contacted force between catheter and blood vessel surgeons can judge whether they can keep inserting the catheter [16–21]. We used an optical fiber sensor to measure this force. Optical fiber is put through the catheter. Sometimes the optical fiber can be used as a guide wire of the catheter.

A micro optical fiber sensor was used to measure front end force of the catheter, meanwhile, the optical fiber sensor served as the guide wire to lead the catheter for inserting and rotating. Figure 10.9 shows the optical fiber sensor we used in this research. FOP-M optical fiber force sensor of FISO Technologies Inc. The sensor is used for pressure measurement. The cross-sectional area was known, so the force could be obtained. The measuring range of the optical fiber force sensor is from 0 to 1 N.

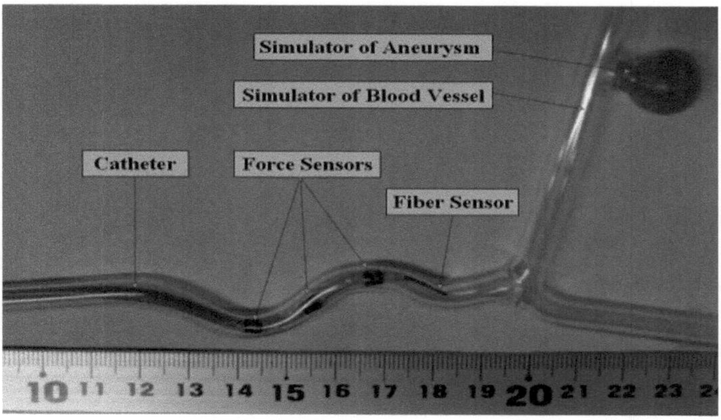

Fig. 10.8 Developed force sensors

Fig. 10.9 Optical fiber sensor

10.3.3 Inserting Experiment in Vitro

A simulation experiment was carried out in vitro. In this paper, we locate two parts of the system in out laboratory; communication between the controller and the catheter manipulator was realized by a LAN communication, time delay was about 10 ms. A rubber tube was used to imitate a blood vessel which has a branch and an aneurysm. The target of the operation is the aneurysm. A catheter should be driven to the target then break it. Figure 10.10 shows the process of the simulation. At the beginning the catheter was inserted into the blood vessel. When the catheter got to the corner as shown in Fig. 10.10a, the catheter should turn to the target but it missed. So the catheter returned and rolled then turned to the right direction as shown in Fig. 10.10b, c. Finally, the catheter reached the target (Fig. 10.10d).

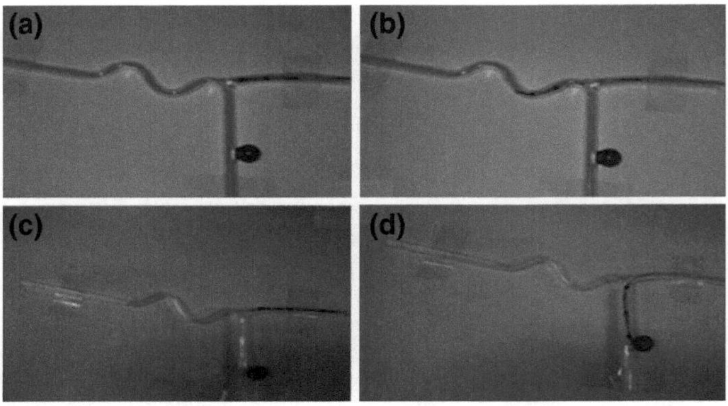

Fig. 10.10 Process of the simulation

10.4 Performance Evaluation of the Developed System

10.4.1 Evaluation of the Controller

To evaluate the measurement the precision of the controller of the system, two experiments were carried out. One is to evaluate the measurement precision of axial movement and the other one for radial movement. For the first experiment, we pulled and pushed the handle to make the movement stage go forward and backward for about 2 min, then measured the axial displacement by a laser sensor (KEYENCE Inc., LK-500, high precision mode, 10μm/mV). A/D convert board (Interface Inc., PCI3329) was applied to get the displacement data to the computer (HP, z400); the sampling frequency was 100 Hz. And at the same time, the measurement data of the controller was sent to the same computer by serial port, the sampling frequency of the controller was set to 100 Hz, baud rate of the serial port was set to 19200. After comparing these two groups of data, the axial measurement precision of the controller could be evaluated. In the second experiment, radial measurement precision is evaluated. Also in the first experiment, we rotated the handle in clockwise and anticlockwise for 2 min, and then measured the rotation angle by outside sensor and by the encoder couple to the handle of controller. A three-axis inertial sensor (Xsens Inc. MTx, resolution 0.1) is fixed on the handle to get the rotation angle. The sampling frequency of the inertial sensor is set to 100 Hz same as the controller. Sampling data of the controller were sent to the computer by serial port.

10.4.2 Evaluation of the Catheter Manipulator

To evaluate the measurement and the precision of the catheter manipulator part, there are two experiments. The first experiment is to evaluate the axial movement precision. The catheter manipulator was programmed to move along a reciprocating trace

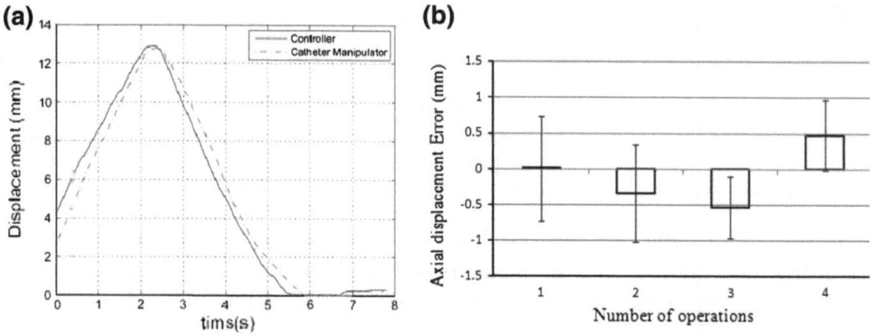

Fig. 10.11 Axial displacement of master and slave side. **a** Axial tracking curve. **b** Axial tracking statistics

in axial. Then the actual displacement of the machine is kept by the laser sensor. The second experiment is to evaluate the radial precision. We programmed the machine to make the catheter frame rotate back and forth. The actual rotation angle is measured by the inertial sensor. In these two experiments, actual data and theory data were compared to obtain the precision of the catheter manipulator.

10.4.3 Evaluation of the Tracking Performance

This experiment was carried out with a local network, the time delay is less than 10 ms. Therefore, we did not consider the time delay between the controller and catheter manipulator. We just evaluated the tracking performance of the axial displacement and rotation. First, we pulled and pushed the controller from right end to left end then went back; we kept the axial displacement by laser sensor in both controller side and in catheter manipulator side. To compare these two groups of data, the synchronization of the controller could be obtained. This procedure was carried out for four times. Similarly, the synchronization of rotation could be obtained.

10.4.4 Experimental Results

1. Evaluation of the Controller and Catheter Manipulator

In controller side, mean value and variance of the error between the laser sensor and the controller are calculated. Similarly, we got mean value and variance of the error between the actual value and the theory. The result is listed in Table 10.1.

2. Evaluation of the Synchronization

Table 10.1 Evaluation of the accuracy and precision

The controller		Accuracy	Precision
The catheter manipulator	Axial (mm)	0.35	±0.025
	Radial (degree)	3.1	±0.5
	Axial (mm)	0.23	±0.04
	Radial (degree)	2.2	±1.5

In this experiment, we measured the axial displacement and rotation of both controller side and catheter manipulator for four times. Figure 10.11 shows the axial tracking experimental results. Figure 10.7a is one of the tracking curve, (b) is the statistics results of four operations. Similarly, Fig. 10.12 shows the rotation tracking experimental results.

Experiments are carried out to evaluate the performance of the robotic catheter manipulation system. The first experimental results listed in Table 10.1 show the precision of the system. The controller gets the input of operator, the measurement precision of the controller is 0.35 mm, and it has a very small variance. Rotation precision of the controller is 3.1° with a variance of ±0.5°. In the catheter manipulator side, the precision of axial direction is 0.23 mm, the rotation precision is 2.2° with a variance of ±1.5°. From the results we can find that the precision in axial direction is better than the radial direction for both sides.

10.4.5 Discussion

Figure 10.11 shows the evaluation of the synchronization of axial direction. In these two experiments, we did not consider the time delay of the remote control. The controller and the catheter manipulator are controlled by DSPs, DSPs communicate with each other with the serial port and a LAN network. Based on the numbers of communication data and the distance, the local network communication is fast enough to make the lag below 1 ms in theory. However, from Fig. 10.11a we can find the movement curve of the controller is faster than the catheter manipulator. It means that the lag exists. And the lag time will be measured in the future. We repeated the procedure for four times. Figure 10.11b shows a statistics results. To evaluate the tracking precision, we have to align the data of two sides. The peak values were aligned artificially to eliminate the effect of the lag. The movement value is about 20 ms. From the results in Fig. 10.11b, the axial error between the controller and the catheter manipulator are below 0.6 mm. The standard deviation is below 1 mm. Figure 10.12 shows the rotation tracking error. Also in the controller part, lag could be found in Fig. 10.12a. Figure 10.12b shows the statistics of the experimental results. Before we obtained the result the data were aligned artificially. The mean error of the rotation is below 1.5°, but with a large standard deviation. It means a not stable radial motion. The large overshoot is mainly caused by the electromagnetic interference to the AD convertor which is used to drive the motor. The control circuit should be redesigned, and with different operator the radial tracking error is different because of the different rotation speed [12].

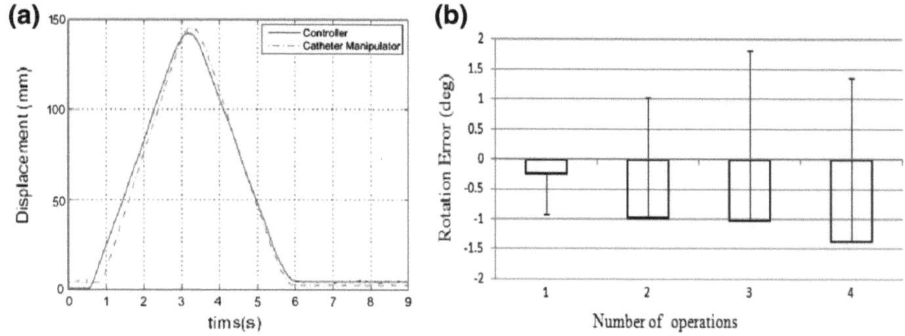

Fig. 10.12 Rotation displacement of master and slave side. **a** Rotation tracking curve. **b** Rotation tracking statistics

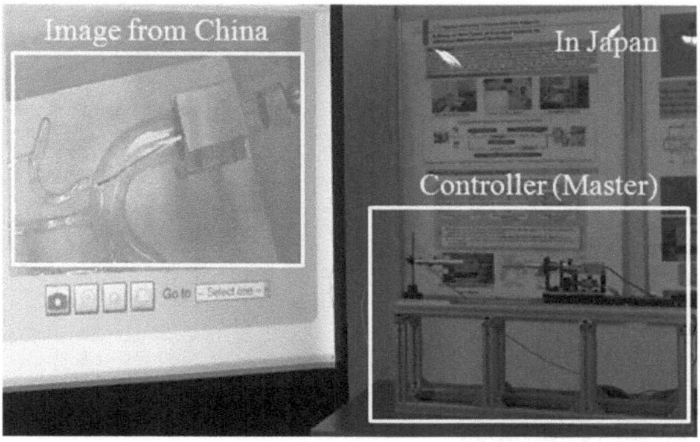

Fig. 10.13 Experimental system (master machine in Japan)

10.4.6 Remote Operation Experiments

Two remote control experiments were carried out. In these experiments, two sets of the developed system were used. Each set includes one controller (master) and one catheter manipulator (slave). One set was placed in Beijing, China and the other one was placed in Japan. An optical fiber sensor was used to monitor the contacted force between the catheter and the blood vessel.

1. Master in Japan

Figure 10.13 shows the experiment system. In this experiment, the controller of the system (master) is placed in Takamatsu, Japan, and the catheter manipulator (slave) is placed in Beijing, China. The server was built in the master side (Japan). The user could see the position of the catheter from the screen. In this experiment, our target was to insert the catheter to a goal position. The displacement of the controller and the catheter manipulator are kept. Contact force was also kept. Figure 10.14 shows the position tracking trajectory of the axial direction. From the experimental result it could be found, the slave can follow the master very well. However, because of the time delay, errors between the two sides are generated.

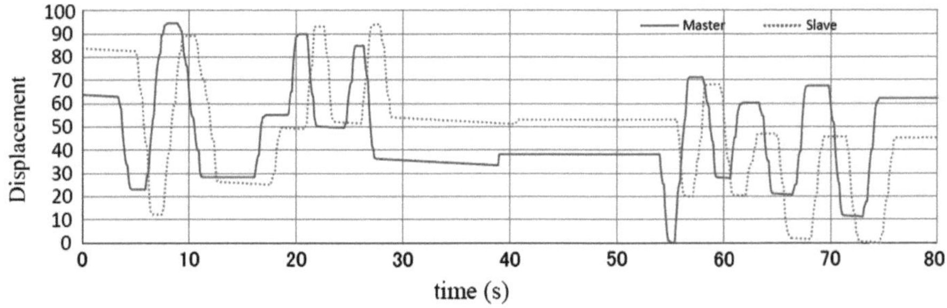

Fig. 10.14 Tracking trajectory of the axial direction

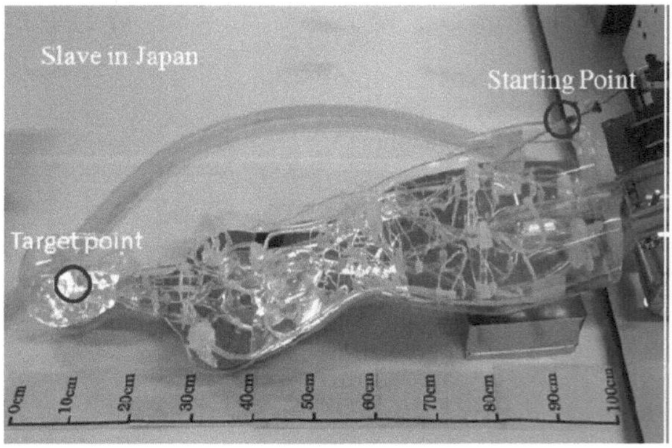

Fig. 10.15 Remote operation of the experiment with EVE model

2. Slave in Japan

Figure 10.15 shows the second experiment. The catheter manipulator of the system (slave) is placed in Takamatsu, Japan, and the controller (master) is placed in Beijing, China.

In this experiment, a human body model EVE is used. The target was inserted the catheter from the leg to the brain with the developed system. Similar to the last experiment, the server was built in Takamatsu, Japan. Figure 10.16 shows the position tracking trajectory of the axial direction.

10.5 Summary

We proposed a robotic catheter robotic manipulator. The system contains two parts, the controller and the catheter manipulator. These two parts have the same movement motion. With this kind of design, the operating procedure becomes visualized. The operating motion is similar to the actual motion of the surgeon's hand. At the same time, with this structure, mechanical feedback could be realized. We carried out experiment to evaluate the performance of the system. Both axial and radial

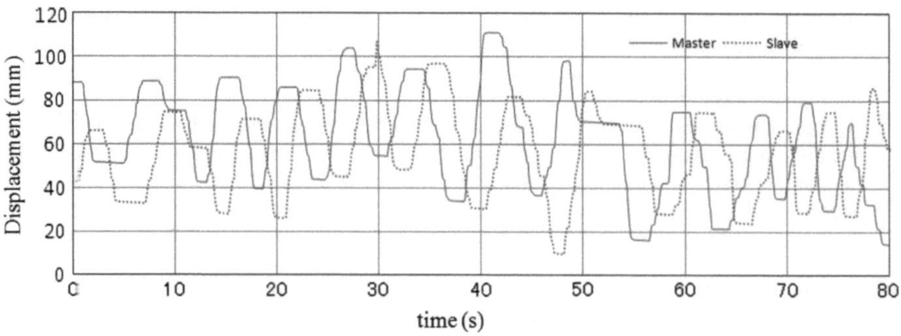

Fig. 10.16 Tracking trajectory of the axial direction

movement precision of the controller and the catheter manipulator are calculated. The measurement precision of the controller can reach 0.35 and 3.1°. The movement precision of the catheter manipulator could reach 0.2 and 2.2°. The tracking precision between the controller and the catheter manipulator are evaluated. When the remote control is used, the error of the tracking is below 0.6 and 1.5°. In order to ensure the safety of surgery, force monitoring system was developed; two kinds of force sensors were installed on the catheter and an in vitro experiment was carried out. The results show that the sensor could measure force information of the surgery in real time. An Internet-based two-way remote control was built between the controller and the manipulator. Experiments were carried out between China and Japan. Experimental results indicate that the two-way operation was realized successfully.

Acknowledgments This research is supported by Kagawa University Characteristic Prior Research fund 2011.

References

1. Kanagaratnam P, Koa-Wing M, Wallace DT, Goldenberg AS, Peters NS, Davies DW (2008) Experience of robotic catheter ablation in humans using a novel remotely steerable catheter sheath. J Interv Cardiac Electrophys 21:19–26
2. Chun JKR, Ernst S, Matthews S, Schmidt B, Bansch D, Boczor S et al (2007) Remote-controlled catheter ablation of accessory pathways: results from the magnetic laboratory. Eur Heart J 28(2):190–195
3. Pappone C, Vicedomini G, Manguso F, Gugliotta F, Mazzone P, Gulletta S, Sora N, Sala S, Marzi A, Augello G, Livolsi L, Santagostino A, Santinelli V (2006) Robotic magnetic navigation for atrial fibrillation ablation. J Am Coll Cardiol 47:1390–1400
4. Willems S, Steven D, Servatius H, Hoffmann BA, Drewitz I, Mullerleile K, Aydin MA, Wegscheider K, Salukhe TV, Meinertz T (2010) Rostock, persistence of pulmonary vein isolation after robotic remote-navigated ablation for atrial fibrillation and its relation to clinical outcome. J Interv Cardiac Electrophys 21:1079–1084
5. http://www.stargen.eu/products/niobe/
6. Saliba W, Reddy VY, Wazni O, Cummings JE et al (2008) Atrial fibrillation ablation using a robotic catheter remote control system: initial human experience and long-term follow-up results. J Am Coll Cardiol 51:2407–2411

7. Ikeda S, Arai F, Fukuda T, Negoro M, Irie K, I Takahashi et al (2005) In vitro patient-tailored anatomaical model of cerebral artery for evaluating medical robots and systems for intravascular neurosurgery. In: Proceedings of IEEE/RSJ international conference on intelligent robots and systems, pp 1558–1563
8. F Arai, R Fujimura, T Fukuda, M Negoro (2002) New catheter driving method using linear stepping mechanism for intravascular neurosurgery. In: Proceedings of the 2002 IEEE international conference on robotics and automation, pp 2944–2949
9. Tercero C, Ikeda S, Uchiyama T et al (2007) Autonomous catheter insertion system using magnetic motion capture sensor for endovascular surgery. Int J Med Robot Comput Assist Surg 3(1):52–58
10. Ramcharitar S, Patterson MS, Geuns RJ et al (2008) Technology insight: magnetic navigation in coronary interventions. Nat Clin Pract Cardiovasc Med 5:148–156
11. Fu Y, Liu H, Wang S et al (2009) Skeleton based active catheter navigation. Int J Med Robot Comput Assist Surg 5(2):125–135
12. Yogesh T, SB Jeffrey, David WH, Drangova M (2009) Design and performance evaluation of a remote catheter navigation system. IEEE Trans Biomed Eng 56(7):1901–1908
13. Puangmali P, Althoefer K, Seneviratne LD, Murphy D, Dasgupta P (2008) State-of-the-art in force and tactile sensing for minimally invasive surgery. IEEE Sens J 8(4):371–381
14. Sedaghati R, Dargahi J, Singh H (2005) Design and modeling of an endoscopic piezoelectric tactile sensor. Int J Solids Struct 42(21–22):5872–5886
15. Takashima K, Shimomura R, Kitou T, Terada H, Yoshinaka K, Ikeuchi K (2007) Contact and friction between catheter and blood vessel. Tribol Int 40:319–328
16. Fu, Y, Gao, A, Liu H, Guo S (2011) The master–slave catheterisation system for positioning the steerable catheter. Int J Mechatron Autom 1(3/4):143–152
17. Wang J, Guo S, Kondo H, Guo J, Tamiya T (2008) A novel catheter operating system with force feedback for medical applications. Int J Inf Acquis 5:83–91
18. N Xiao, S Guo, J Guo, X Xiao, T Tamiya (2011) Development of a kind of robotic catheter manipulation system. In: Proceedings of the 2011 ieee international conference on robotics and biomimetics, Phuket, pp 32–37, 7–11 Dec
19. Guo J, Guo S, Xiao N, Ma X, Yoshida S, Tamiya T, Kawanishi M (2012) A novel robotic catheter system with force and visual feedback for vascular interventional surgery. Int J Mechatron Autom 2(1):15–24
20. N Xiao, J Guo, S Guo, T Tamiya (2012) A robotic catheter system with real-time force feedback and monitor. J Australas Phys Eng Sci Med 4, July (Online)
21. Wu YC, Chen JS (2011) Toward the identification of EMG-signal and its bio-feedback application. Int J Mechatron Autom 1(2):112–120

Index

Y. Guo (ed.), *Selected Topics in Micro/Nano-Robotics for Biomedical Applications*,
DOI: 10.1007/978-1-4419-8411-1,

MIX
Papier aus verantwortungsvollen Quellen
Paper from responsible sources
FSC® C105338

If you have any concerns about our products,
you can contact us on
ProductSafety@springernature.com

In case Publisher is established outside the EU,
the EU authorized representative is:
Springer Nature Customer Service Center GmbH
Europaplatz 3, 69115 Heidelberg, Germany

Printed by Libri Plureos GmbH
in Hamburg, Germany